R0002714192

CHICAGO PUBLIC LIBRARY
HAROLD WASHINGTON LIBRARY CENTER

R0002714192

TJ   DeVito, Alfred.
163.2
.D48   Activities handbook
         for energy
         education

TJ   DeVito, Alfred.
163.2
.D48   Activities handbook
         for energy
         education

R00027 14192

DATE	BORROWER'S NAME	
NOV 7 1984	Hanley Park	

Business/Science/Technology
Division

© THE BAKER & TAYLOR CO

ACTIVITIES HANDBOOK FOR ENERGY EDUCATION

**OTHER GOODYEAR BOOKS
IN SCIENCE, MATH, & SOCIAL STUDIES**

THE CHALLENGE OF TEACHING SOCIAL STUDIES IN THE ELEMENTARY SCHOOL, 3rd ed.
Dorothy J. Skeel

DR. JIM'S ELEMENTARY MATH PRESCRIPTIONS
James L. Overholt

THE EARTHPEOPLE ACTIVITY BOOK People, Places, Pleasures and Other Delights
Joe Abruscato and Jack Hassard

ECONOMY SIZE From Barter to Business with Ideas, Activities, and Poems
Carol Katzman and Joyce King

LEARNING TO THINK AND CHOOSE Decision-Making Episodes for the Middle Grades
J. Doyle Casteel

LOVING AND BEYOND Science Teaching for the Humanistic Classroom
Joe Abruscato and Jack Hassard

MATH ACTIVITIES WITH A PORPOISE
Runelle Konsler and Lauren Mirabella

MATHMATTERS
Randall Souviney, Tamara Keyser, Alan Sarver

MEASUREMENT AND THE CHILD'S ENVIRONMENT
Tamara Keyser and Randall Souviney

MULTICULTURAL SPOKEN HERE
Discovering America's People Through Language Arts and Library Skills
Josephine Chase and Linda Parth

THE OTHER SIDE OF THE REPORT CARD A How-to-Do-It Program for Affective Education
Larry Chase

SELF-SCIENCE The Subject is Me
Karen F. Stone and Harold Q. Dillehunt

TEACHING ETHNIC AWARENESS Methods and Materials for the Elementary Classroom
Edith King

THE WHOLE COSMOS CATALOG OF SCIENCE ACTIVITIES For Kids of All Ages
Joe Abruscato and Jack Hassard

For information about these, or Goodyear books in Language Arts, Reading,
General Methods, and Centers, write to

> Janet Jackson
> *Goodyear Publishing Company*
> *1640 Fifth Street*
> *Santa Monica, CA 90401*
> (213) 393-6731

ACTIVITIES HANDBOOK
FOR
ENERGY EDUCATION

Alfred De Vito,
Professor of Education

Gerald H. Krockover,
Professor of Education and Geosciences
Purdue University

Goodyear Publishing Company, Inc.
Santa Monica, California

Library of Congress Cataloging in Publication Data

DeVito, Alfred.
  Activities handbook for energy education.

    1. Power resources — Study and teaching.
I. Krockover, Gerald H., joint author. II. Title.
TJ163.2.D48   333.79'07'1   81-384
ISBN 0-8302-0034-7   AACR1

TJ
163.2
.D48

Copyright © 1981 by Goodyear Publishing Company, Inc.
Santa Monica, California 90401

All rights reserved. No part of this book may be reproduced
in any form or by any means
without permission in writing from the publisher.

Current printing (last digit):
10 9 8 7 6 5 4 3 2 1

ISBN: 0-8302-0034-7
Y-0034-2

Cover and text design by Karen McBride
Illustrations by Karen McBride and Madeleine Moore
Printed in the United States of America

BST

Contents

Introduction

The *Activities Handbook for Energy Education* was designed to provide educators with a spectral view of energy in terms of past, present, and future potential energy practices. Much has been written about energy and the current energy crisis. Little has been done to assemble this plethora of information into a readily assimilated form for educators and children. The primary premise of this handbook is to present energy knowledge and to translate this knowledge into suitable activities interpretable by children. Thus, this handbook could assist educators in developing their own energy education program. To augment this development, over 140 ideas and activities have been included to assist educators in the development of energy education materials for individual situations.

This handbook does not address itself to the role of social values in making energy decisions. It does not deal with the politics of energy use. It does not deal with altruistic, moral, or economic persuasion as solutions to the energy crisis. This is not to say that these topics are unimportant. These topics are rooted in the foundational energy knowledge presented in this handbook. Coverage of these topics by educators could emerge from the energy knowledge that they garner from this handbook.

It was concluded that space plus the complexity of these issues dictated that these topics could not be adequately treated within the confines of this handbook. Rather than cloud issues, our efforts were confined to providing general background information relative to energy and including action involvement for children through the selection and suggestion of appropriate energy activities.

Chapter 1, "Energy: A Delicate Dilemma," presents energy problems as they exist today with activities designed to provide an introduction to energy and energy usage. Vignettes are also included to illustrate the fact that the answer to our energy concerns is not a "yes, no, or either/or" situation. A glossary of energy terms along with an introduction to the concept of energy is also included.

Chapter 2, "What Are the Sources of Energy?" provides background and activities for the major energy sources used in the past, used today, and anticipated use in the future. Sources such as solar, wood, coal, steam, oil, radiant, wind, gas, and nuclear are included. Ideas and activities are provided to introduce energy sources from each of the source areas.

Chapter 3, "What Are the Uses of Energy?" provides a background pertaining to energy uses in transportation, household and commercial, industry, and electric power generation. Ideas and activities are provided to illuminate the uses of energy in our society.

Chapter 4, "How Does One Conserve Energy?" presents the rationale, background, ideas, and activities for energy conservation through the use of appliances, insulation, transportation, and recycling efforts.

Chapter 5, "What Will Our Energy Future Be Like?" presents the background, ideas and activities related to energy use in the twenty-first century. Included in this chapter are descriptions of energy sources such as tides, geothermal, nuclear fusion, wind, ocean thermal energy conversion, fuel cells, hydrogen fuel, solar, solid and organic waste, oil shale, tar sands, and hydro energy.

The bibliography includes a comprehensive list of over 100 additional sources of ideas and activities for energy education. Many of the materials listed in this section are free or inexpensive and can be incorporated into a variety of grade levels and curricular areas.

Dispersed throughout the book are more than seventy "Did You Know?" and more than twenty-five "Does It Make Sense?" or "The Answer Is" ruminations. These can be used directly or modified to serve as puzzler type activities or points to ponder. They should serve to promote discussion and further research into the topic they relate to in the chapter.

The "Did You Know?" that follows can serve as an example.

DID YOU KNOW?

"There is one obstacle to further advance . . . The increasing price of fuels necessary to work machinery. Coal and oil are going up and are strictly limited in quantity . . . We are spendthrifts in the matter of fuel and are using our capital for our running expenses."

"In relation to coal and oil, the world's annual consumption has become so enormous that we are now actually within measurable distance of the end of the supply. What shall we do when we have no more coal or oil?"

"Alcohol makes a beautifully clean and efficient fuel . . . We can make alcohol from sawdust, the waste product of our mills . . . from cornstalks and in fact from almost any vegetable matter capable of fermentation. Our growing crops and even weeds can be used. The waste products of our farms are available for this purpose, and even the garbage from our cities."

Alexander Graham Bell
February 1, 1917

Questions that one might ask include:

1. What have other famous people of the past had to say about our energy future? Find out.
2. Why didn't we listen to Alexander Graham Bell's recommendation in 1917?
3. Find out how alcohol can be made from sawdust, cornstalks, or other crops. Why aren't we doing this?

Activities and ideas are found throughout each chapter. Some are found at the end of a chapter. They are included as appropriate reinforcement for the material being presented.

We hope that as a result of your involvement with the materials and activities in this book, energy education will permeate all of the subjects in your classroom, all of the classrooms in your school, and all of the schools in your community. The end result will be a new generation of responsible energy education individuals. This is the ultimate answer to the energy crisis that our country faces today.

The minds of the modern inventors, the skilled scientists, the creative architects, the futuristic farmers, and the knowledgeable consumers will all be needed to solve the energy problems that we face. Out of the hard shell of the energy crisis of today will hatch the new energy forms of tomorrow.

Acknowledgments

The authors would like to acknowledge Christopher Jennison and Nancy Carter of Goodyear Publishing Company for their efforts in assuring a successful energy education book.

This book is dedicated to the proposition that the energy problems we face are solvable and will be solved in the future.

Alfred De Vito
Gerald H. Krockover

I believe   that man will not merely endure
                  he will prevail

     He is immortal
not because he alone among creatures   has an inexhaustible voice
       but because he has a soul
   a spirit capable of compassion
             and sacrifice and endurance
     WILLIAM FAULKNER

Energy: A Delicate Dilemma

"One of the greatest failures of national leadership in recent history is the failure to convince the American people of the urgency of our energy problems."

Jimmy Carter
39th President of the United States

Energy consumption in the United States is about 10,000 watts per person per day. This energy is generated by many different means, and is used in transportation; heating and air conditioning our homes, offices, stores, and factories; and in the operation of the industrial processes in the economy. Today the United States spends about $27 billion per year, or $215 per person, for imported oil, as compared to 1970s expenses of about $3 billion, or $15 per person.

The sun has been, and continues to be, the source of all energy since the world began. All living things depend on it for life. Early man got energy from the sun by eating the food that the sun helped to grow. As people looked for ways to accomplish work other than through their own muscle and/or animal power, they discovered and began to use alternative sources of energy—the wind, water, steam, coal, oil, gas, and nuclear. We continue to search for sources of energy to do work, but we are approaching a delicate dilemma. What energy sources will be best for our future? What shall our energy development efforts be directed toward? Which way should we turn?

1

Who Is Right? Vignettes of Energy Conflicts

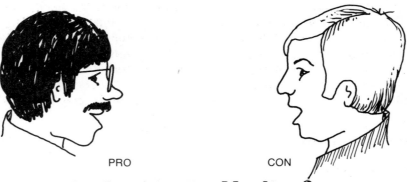

PRO   CON

Is the Answer Nuclear?

PRO

In 1985, there will be approximately 175 nuclear power reactors in the United States and a potential of 650 reactors available by the end of the century. In the year 2000, nuclear power could be supplying energy equal to that of 2 billion tons of coal per year. A few of the many advantages of nuclear power as an energy source follow.

• The United States has domestic uranium resources vast enough to support a huge nuclear power industry for many centuries.

• Uranium mining is environmentally preferable to strip-mining less energy intensive fuels such as coal or oil shale.

• One uranium miner's daily output equals that of twenty-six coal miners.

• A single truckload of refined uranium holds energy equal to that contained in four thousand cars, full of 100 tons of coal.

• Nuclear power is a proven technology that generates electric power at economical rates.

• Nuclear fuel can be stockpiled to keep several years supply in inventory at a plant; thus eliminating the risk of a strike.

• Nuclear power is safe. There has never been a single commercial accident that has been injurious to anyone offsite.

• The disposal of radioactive

CON

Day in and day out, the steam flows from the tower that cools the heat of one of the nation's largest nuclear power plants. The plume of steam rises so high that it can be seen for 60 miles in any direction. The cooling tower is over 500 feet (150 meters) high. Now, for the first time since the use of nuclear power as an energy source, we are faced with the issue of disposal of nuclear waste. What do we do with the first load of spent radioactive fuel that's about to be discharged from the plant?

Everyone understands what garbage is, but this garbage is different—it contains the elements strontium and cesium. Both of these elements emit gamma rays (which can penetrate lead) and both stay radioactive for generations, meaning they will still be generating gamma rays over one thousand years from now.

What to do with radioactive waste is the most difficult energy question facing the United States. As of 1978, the United States had already accumulated over 5,000 tons of spent nuclear fuel. This figure doesn't include the 2 million tons of radioactive clothing and medicines discarded and buried underground in the last twenty-five years. It also doesn't include the 100 million gallons of hot waste generated by the nuclear weapons program.

wastes is a manageable problem, involving comparatively small volumes of solid wastes.

• The operation of nuclear power plants is environmentally clean; the very small quantities of radioactive releases are carefully monitored and present no danger to community health.

Increased dependence on nuclear power plants by the United States would enlarge the spent fuel figure to 10,000 tons by 1985, and over 100,000 tons by the year 2000.

The states of Oregon, Michigan, and California have already enacted laws banning the "permanent" disposal of radioactive wastes in their states.

Is the Answer Nuclear?

What Do You Think?

Is the Answer Coal?

PRO

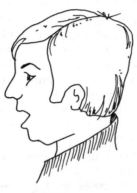

CON

In spite of our growing dependence on other sources of power, fossil fuels will still be our main source of energy through the year 2000. Our immediate need is to derive maximum energy from coal while preserving environmental quality. We have at least a two hundred year supply of clean and accessible coal. Power companies and industries must shift to this source of energy, and we must invest in improved mining efficiency, cleaner combustion technology, and a better transportation system for moving coal to its users. The shift from oil and gas usage to coal will reduce the nation's dependence on an insecure foreign energy supply. In the future coal may be transported directly to the power plant from the mine through pipelines as a slurry.

Coal is not easy to mine; it is not easy to clean up after; it is not easy to transport. Its wastes are unaesthetic and damaging to one's health. Deep mines are costly to companies and to the lungs of miners. Strip mines are costly to the environment and to the mining companies required to rehabilitate the land they have ripped up — putting back the rocks, restoring contours, putting back a blanket of topsoil, seeding, and planting are cost and time intensive activities. Restoration of the landscape is particularly difficult in the semi-arid regions of the West where much of the nation's untapped coal lies buried. A major part of the capital investment and maintenance costs of coal-fired generating plants is the cost of protecting the quality of air and water in the vicinity of the plant. The effec-

We must take steps to increase the use of our nation's vast supply of coal. We need to make some reasonable compromises between energy and the environment; we need to relax some of the rigid and utopian environmental standards imposed in an era of wishful thinking; and we need to provide an economic and political atmosphere which will encourage the long-term development of our nation's coal resources.

tiveness of the present environmental protection systems is questionable. There is also evidence that the techniques required by the Environmental Protection Agency (EPA) regulations to clean smokestack emissions may actually create pollutants of another type. We must also remember that the "best" fossil fuel plants that we now have can convert only about 40% of the energy stored in these fuels into electricity with about 60% lost as heat.

The disadvantages of coal mining include: stream pollution, floods, landslides, sedimentation, loss of fish and wildlife habitats, sulphuric acid pollution, and thick yellow mining wastes.

Is the Answer Coal?

What Do You Think?

Is the Answer Solar?

PRO

CON

Using the direct radiant energy of the sun to heat water for household uses and to warm living and working spaces (which constitute more than half our domestic demands) is practical and real *right now*. There is a growing variety of solar collectors of varying complexity and cost available for both commercial and private use. The most significant use of solar energy in the near future will be for home consumption (heating and

The sun is not free. Its generous radiation is paid for in advance with the high capital costs of solar installation, backup, and storage systems. Electricity produced by solar thermal plants utilizing space-age technology will cost hundreds of dollars a watt instead of fractions of cents. Reducing the cost will require a major breakthrough comparable to the development of the transistor and miniaturized circuitry. Unfortu-

cooling). Solar thermal plants that produce heat equivalent to that of several thousand suns must be developed and built. This heat can be used to vaporize a liquid which can then be used to run a conventional generating turbine. A major breakthrough in the cost of solar energy generation will come soon.

Remember how large and expensive calculators once were and how rapidly they have become both pocket-size and inexpensive? Solar energy costs will follow the trend of pocket calculators in terms of price, usefulness and benefits to society.

nately, breakthroughs do not always conform to our schedules. The sun, for all its power and versatility, has a major flaw; its rays reach us only half the time under the best of circumstances and there are frequent outages due to passing clouds, smog and fog. Several days of dense overcast skies could be disastrous. To insure a steady supply of power, electric companies will have to keep conventional plants for backup services. Even the most optimistic experts who foresee that solar energy in its various manifestations will provide nearly all of our energy needs one hundred years from now expect no more than an 8% contribution by the year 2000. The only realistic solar collector available to an individual in the near future will be his south window.

Is the Answer Solar?

What Do You Think?

Now that you've read some of our energy conflict vignettes, why don't you, as Pro or Con, try to develop some of your own vignettes in response to the following questions.

Is the answer geothermal?

Is the answer wind?

Is the answer water?

Is the answer thermal?

Next, assume one of the roles assigned to Pro or Con in the following list and address the issues from that point of view.

Pro as the	*Con as the*
parent	child
oil company person	conservationist
oil company person	stockholder
scientist	economist
economist	scientist
consumer	business person
business person	consumer
electric utility	consumer
regulatory agency	electric utility

Can you think of other roles that you might like to assume?

We have not yet reached a plateau on the growth curve of energy use and energy consumption should continue to increase during the next fifty to one hundred years. Our choice of an energy source for the future will be crucial to those who live on this earth in the twenty-first century and beyond.

DOES IT MAKE SENSE? RESERVES VS CONSUMPTION?

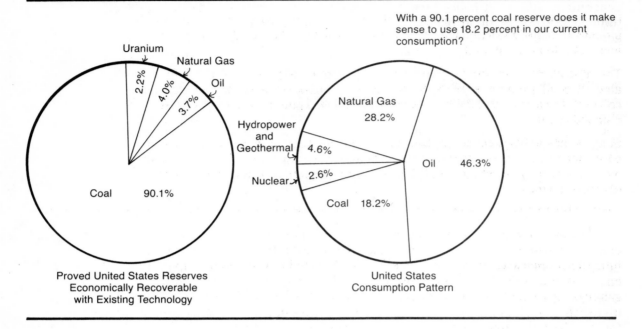

With a 90.1 percent coal reserve does it make sense to use 18.2 percent in our current consumption?

Uranium 2.2%
Natural Gas 4.0%
Oil 3.7%
Coal 90.1%

Proved United States Reserves
Economically Recoverable
with Existing Technology

Natural Gas 28.2%
Hydropower and Geothermal 4.6%
Nuclear 2.6%
Coal 18.2%
Oil 46.3%

United States
Consumption Pattern

Energy: Who Needs It Anyway?

Energy is a pretty independent thing. It can and does exist without us. But we cannot exist without it.

One of the characteristics of living things is mobility. We move. We must move to survive. Sometimes we move voluntarily and other times we move involuntarily. If someone throws a ball to you or at you, you move to catch or dodge the ball. This action or movement is usually a voluntary movement. Breathing, by contrast, is an involuntary movement. We are not always aware that we are breathing. In a deep sleep, you breathe unconscious of the process. In either case we or some part of us is constantly moving. This involves work, and work takes energy.

Work is a subjectively defined word. Something that is work for you may be pleasurable for someone else, and vice versa. In either case, whether you call an action work or pleasure, it takes energy. Does thinking involve work? Does thinking involve energy?

In our natural system, plants are the only energy producers. All other living things, including you, are energy consumers. Plants, unlike humans, directly convert the sun's energy into food. This process is called photosynthesis. All other living things owe their existence in one way or another (directly or indirectly) to plant life. When people or animals eat plants they receive the plant's energy. When people or animals eat animals who live on plants they receive the energy from the animals who received it from the plants. This energy is then converted into other forms of energy by the consumer. The sun's energy and plants' ability to convert the sun's energy into food are the first step in the long cycle of life and energy as we know it. How long could we survive without plants; without the sun?

DID YOU KNOW?

Light is one of the most important forms of radiant energy for living organisms. All life processes ultimately depend upon the ability of green plants to capture light energy and use it to produce food.

The amount of sugar that a green plant produces is directly proportional to the amount of light energy supplied to the leaves. Chlorophyll absorbs sunlight and converts it to a form of chemical energy. The chemical energy can then be used to produce sugar.

Since photosynthesis consumes less than 13% of the solar energy that reaches the earth, and about 28% is sufficient to operate the earth's heat machine, almost 72% of our solar energy or about 180,000 calories per square centimeter per year is returned to space.

Human activity or work can be measured in kilowatt hours (kWh) of energy. One hour of normal activity equals .1 kWh (or 100 watts). In one year a human's normal activity can use 876 kWh. If a human being could pull up to an energy pump at the local gas station, he would need to fill up with at least 24 gallons of gasoline to get sufficient energy to equal 876 kWh. This would equip him with enough energy to do normal daily activities only. Someone training for the Olympics would require more gallons to generate more kWh.

DID YOU KNOW?

Each day, the average American uses energy equivalent to:

    13.6 pounds of coal

     3.3 gallons of oil

    297.0 cubic feet of natural gas

     3.7 kWh of hydro-electric power, and

      .7 kWh of nuclear power

Human beings do not run on gasoline. We need air, water, and food to stay alive. Another basic life requirement is the energy needed for the body to process the necessary air, water, and food. It takes energy to fill your lungs with air and to empty them. Water, present in and necessary to all organic tissues, is consumed and used in a variety of fluid forms. It takes energy to take water in, to use it, and then to expel the fluid secreted by the kidneys. It takes energy to eat food. In turn, the food plus air and water gives us energy to survive. The heart, a human pump, does work. Work requires energy. Survival is a balance of input (air, water, and food) and output (air, water, and food). It takes energy to get energy.

Food is an energy source. People eat a chocolate candy bar for "quick" energy. Then, they worry about the calories. Someone is always worried about taking in too many calories. The calorie values assigned to foods are equal to the energy released when the foods are burned in the cells of your body. A calorie is defined as the amount of heat needed to change the temperature of 1 gram of water to 1 degree Celsius. The calorie is a unit used to measure heat content.

A gram calorie is referred to as a small calorie (c). A kilogram calorie or a large calorie (C) is the quantity of heat equal to 1000 gram calories. Fuel values of foods are given in terms of large calories (C) or kilocalories.

The daily caloric need for an average physically active man should be 3,000 calories; for the average moderately active woman, 2,400 calories; for an active teenage boy, 4,000 calories; and for an active teenage girl, 2,500 calories.

DID YOU KNOW?

By the time 1 calorie of food reaches your table approximately 9 calories of energy have been used to get it there.

You are what you eat. If you eat more food (energy input) than you expend (energy output), you'll gain weight. Conversely, if you eat less food (energy input) than you expend (energy output), you lose weight. In either case your energy level is affected. Your body needs energy to maintain life activities. Necessary energy is released in the cells as part of respiration. Fats and carbohydrates are oxidized in the cells, releasing the energy needed by the body. Some of this energy is used to maintain a constant body temperature.

FUEL FROM SOME SNACK FOODS

Food	Calories in an average serving
Hamburger on a bun	500
Peanut butter sandwich	370
Potato chips	110
Chocolate bar, small, plain	190
Chocolate bar, small with nuts	275
Malted milk	450

HOW DO WE MEASURE ENERGY?

Scientists make a distinction between heat and temperature. Temperature is the measure of hotness or coldness of an object and its units are in degrees of Fahrenheit, Celsius, or Kelvin. Heat is a measure of the agent that makes things get hot. Heat is measured in calories similar to, but not exactly the same, as the calories you worry about when you go on a diet. It takes 1,000 calories to make 1 diet calorie, called a kilocalorie (C).

One calorie is the amount of heat needed to raise the temperature of 1,000 grams of water 1 degree Celsius. A Calorie is one thousand times the value of a small calorie (c), the heat unit used most often in science. One small calorie (c) is the amount of heat needed to raise the temperature of 1 gram of water 1 degree Celsius.

Obtain a bag of unshelled peanuts. Randomly select one peanut and shell it. Determine the mass of the peanut and record it in the chart below. Obtain a small metal juice can and pour a measured amount of water (in milliliters) into it. Record the initial temperature of the water in degrees Celsius. Hold the peanut in place with a bent paper clip stuck into a cork (see drawing) for

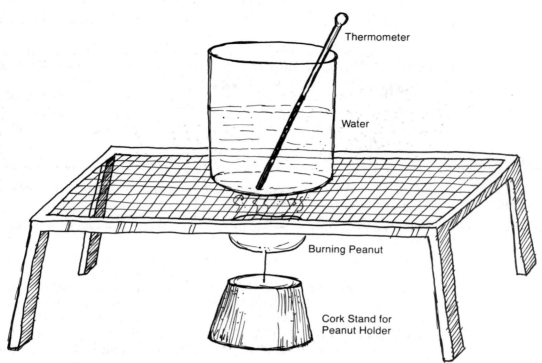

support. Carefully strike a kitchen match and light the peanut. After the peanut has burned completely, record the final temperature of the heated water. Subtract the initial temperature from the final temperature; this is the temperature change.

Volume of water used _____

Determine the temperature of the water_____ (Don't forget to indicate which units of measurement you used.)

Determine the mass of the peanut _____ (Did you take it out of the shell?)

Light the peanut and allow it to heat the water.

What is the temperature of the water after being heated by the peanut?

_____

Record the new temperature of the water. How much did it change?

_____

Using the formula below, calculate the number of calories of heat given off by the peanut to the water.

cal = volume of water (ml) x 1 g/l ml x 1 cal/g-deg x temp. chg °C

Take the volume of water in ml times 1 g/ml (density of water) times 1 cal/g-deg (heat capacity of water) times the temperature change (°C). This enables all your units to cancel except calories, which is what you want!

Now, change your answer to Kilocalories (C). Since it takes 1,000 calories (c) to make 1 Kilocalorie (C), how many Kilocalories of heat were given off by the peanut?

BODY ENERGY PROFILE

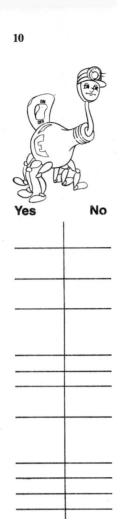

Yes   No

They say everyone can be identified by their fingerprints. Fingerprints are unique to every individual. No two are alike. Some researchers say this is also true for earprints or earshapes.

Along with fingerprints and earshapes, you probably could be identified by your energy profile. Your profile is based on your likes and dislikes, your desire for comfort, and your general appetite for the things that require energy. How would you score on the following questions (answer each question with a *yes* or *no* response). When completed, tally the *yes* responses.

- You prefer taking showers to soaking in a tub.
- When you shower, you love to remain under the water as long as you can, or until all the hot water is gone.
- You hate cool or cold running water for bathing; you need warm to hot running water.
- When you enter the house from the outdoors, you immediately turn on the TV (or radio, stereo, etc.) and then continue on, not watching or listening to the appliance you have turned on.
- You prefer to drive or be driven rather than walk—even short distances.
- You keep all the lights burning even when you're occupying one room.
- You open the refrigerator door and hold it open while you study all the items in it.
- Every time you wash your hands, you run the hot water tap rather than the cold water tap even though the hot water has never reached the sink by the time you have finished washing your hands.
- You like everything well-cooked, well-done, almost burned.
- You like to change your clothes every day.
- You like to change your clothes several times a day.
- You like to wash your hair every day.
- You like to use the hair dryer and blower when you finish washing your hair.
- You like to play the stereo even when you are away from it for extended periods of time.
- You like new gadgets such as electric toothbrushes, electric shoeshiners, electric blankets, etc.
- You own one or more radios.
- You prefer having an air conditioner to just an open window.
- You love fresh air and on cold nights you like to have a window open, even though the heat is on.
- When you drive or are driven in an automobile, you prefer to get where you are going directly.
- You use materials such as soap, water, face tissues, and paper napkins excessively.
- You prefer to use a power mower rather than a hand mower when cutting the grass.
- You would rather use the electric dishwasher than do the dishes by hand.
- You prefer battery operated gadgets such as clocks, games, toys, etc., to those that wind up.
- You prefer to use the plastic kitchen bags for garbage instead of brown, grocery-type bags.

TOTAL

DOES IT MAKE SENSE?

In this country 75% of the cars being operated at one time carry only one driver.

DID YOU KNOW

A 100-watt bulb that burns for one year costs about $30.00 and uses about 1000 lbs. of coal to generate the electricity needed to keep the bulb burning.

WHEN IS ENOUGH, ENOUGH?

You are going camping. Read each of the following options: Which option best describes the equipment that suits you?

- Option 1: *Low energy requirements (bare minimum)*
  knapsack or backpack
  groundcloth
  poncho
  knife
  one change of clothes
  soap
  no cooking utensils

- Option 2: *High energy requirements (maximum)*
  knapsack or backpack
  sleeping mattress
  air mattress
  dry foods
  cooking utensils
  canteen
  compass
  knife
  axe

pedometer	candles
pocket radio	flashlight
camera and film	comic books
first aid packet	playing cards
suntan oil	several changes of clothes
insect repellent	25 feet of rope
tissue paper	marshmallows

• Option 3:   *Super, high-energy requirements*
               (maxi-maximum)

When is enough, enough? The low energy option may be too minimal for you. While the high energy option may contain many "things" that you might like, it would be a physical challenge just to carry all these items. How do you feel about the super, high-energy option?

Simplicity is an essential aspect of energy conservation. Sophisticated variations of simple things require extra energy. Suppose you wanted to get somewhere 5 miles from where you are now. Would you walk? Walking requires energy supplied by you, plus the use of some shoe leather. Or, you could choose to ride a bicycle. This also takes your energy, plus some shoe leather, and a bicycle. The bicycle consists of metal, rubber, grease, and oil. The metal had to be mined, refined, shaped, cut, welded, assembled, transported, stored, displayed, advertised, and then sold to you. We have not mentioned the energy needed to acquire the rubber used for the tires on the bicycle. Nor have we mentioned the energy needed to fabricate the chain and lock that you will need to secure the bicycle when you are not riding it. Each step in the process from raw metal to the finished bicycle required additional energy in a variety of forms. More energy is needed in proportion to the complexity of the object(s) you use to aid yourself. This increase of energy expenditure, created by more elaborate uses of energy to accomplish a task, is like an inverted triangle (▽). Simple, one step operations are at the bottom of the triangle, more complicated ones are near the top.

Very little "waste" oil (dirty or used) is recycled these days. During World War II waste oil was recycled. Recycling oil today could save us over one billion gallons of oil a year.

Bicycles are twenty-eight times more efficient than cars; walking is seventeen times more efficient than cars; buses are four times more efficient than cars; trains are two and a half times more efficient than cars. Only airplanes are less efficient than cars. Why aren't we all riding bicycles?

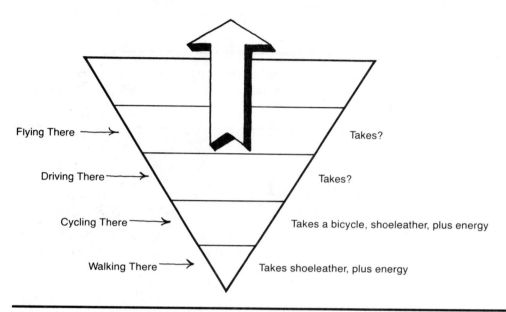

Flying There → Takes?

Driving There → Takes?

Cycling There → Takes a bicycle, shoeleather, plus energy

Walking There → Takes shoeleather, plus energy

The time has come for us to consider how much energy we use to accomplish a specific task and then to ask ourselves if the methods we are now using are the most energy-efficient ones available to us.

How Can We Reduce Our Energy Consumption?

- Use natural things. Using things from our natural surroundings in their natural state conserves energy. For example, minimum energy is used when you eat wild hazel nuts. If you purchase roasted, packaged hazel nuts, the energy level just to create the product is substantially higher. It takes energy to gather the nuts, clean them, roast them, package them, and then transport them.
- Get the maximum use out of things. Do not throw things away that are still useful. When you buy something think of how much use you will be able to get from it over a prolonged period of time. Is the item worth the price in terms of the use and future use of the item?
- Ask yourself, "Does my comfort level demand this?" This is often a difficult question to answer. Is it essential to your well being or not? What price are you willing to pay for conserving energy?

DOES IT MAKE SENSE?

The lifetime energy consumption of a person living in the United States is: 170 tons of coal, 2000 barrels of oil, and 7.5 million cubic feet of natural gas.

DID YOU KNOW?

It takes about 2 fluid ounces of petroleum (or about 1⅓ ounces of coal) to burn a 100-watt light bulb for one hour.

DID YOU KNOW?

The United States does not have many commodities anymore. Our balance of trade suffers from our increasing need to import foreign raw materials. Recycling various materials would help lower our imports.

What Is Energy?

Energy, the most fundamental of all physical concepts, is a force that produces change. It is measured by its effect upon matter. Energy does not take up space. There is no such thing as 3 cubic feet, or 4 pounds 3 ounces of energy. It has no mass, but it is very noticeable. In general, the greater the change or effect created in or on something, the greater the amount of energy used. Energy is the ability to work. All the familiar uses of energy fall into two major categories — the ability to do mechanical work or to produce a change in temperature (to heat or cool).

The work of energy is in evidence all around us. Yet, we rarely wonder about where to find it, to use it, and why it is so important. Energy is needed to keep the earth spinning on its axis. It feels as though the earth is motionless in space. This is not so. If you lived near the equator your speed, as the earth turns on its axis, would be almost 1,000 miles per hour. Simultaneously, the planet Earth is speeding around the sun at about 1.5 million miles per day or 66,000 miles per hour. While you are spinning around on the earth's axis, and the earth is revolving around the sun, you, as part of the solar system, are speeding out into space at an estimated speed of 500 million miles per hour. Who or what is furnishing the energy to accomplish all this work?

DID YOU KNOW?

About half the energy needed to run a city comes straight from nature, in the form of water flows, air currents, and solar energy.

Energy produces changes in the air, the sea, and the land. Energy moves gases, liquids, and solids. Energy is responsible for moving light through space at a velocity of 186,000 miles per second. Energy does a lot of work.

A better understanding of what energy is can be achieved by examining some of the many forms in which it exists.

Kinetic energy is the energy of an object in motion. The water droplets in a rainstorm have kinetic energy.

Potential energy is stored energy. It can be stored energy an object possesses because of its high relative position. A rock balanced on the edge of a cliff has a lot of potential energy because of its high relative position.

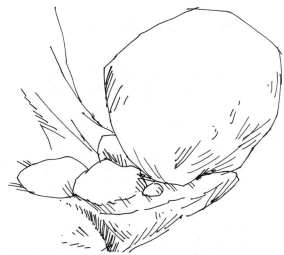

If the balanced rock is no longer in balance and is falling down the face of the cliff, it no longer possesses potential energy. Its energy is transformed to another form of energy. What type of energy does the falling rock now have?

Mechanical energy is energy directly involved with moving matter. A baseball driven over the ball park fence for a homerun is a good example of mechanical or muscular energy.

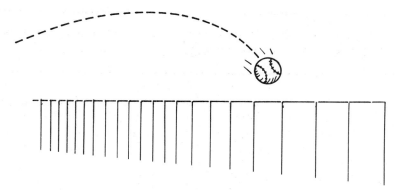

Electrical Energy is energy resulting from the flow of electrons along a conductor. A battery is a means for storing electrical energy.

Chemical energy is the energy possessed by chemical compounds and is the most fundamental form of energy in the life processes. Coal, natural gas, and oil are important examples of materials having a great deal of chemical energy. These are referred to as primary fuels.

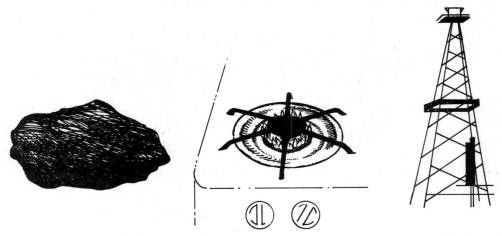

Nuclear energy is the energy associated with the particles in the nucleus of atoms. Large forces exist within the nucleus of the atoms that form the building blocks of all matter. Certain chemical elements have unstable nuclei. These elements undergo constant spontaneous disintegration in an attempt to become more stable products.

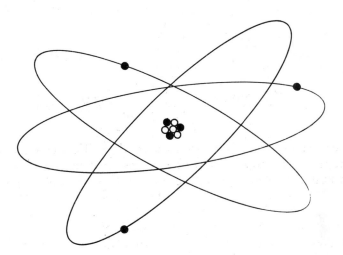

Sound Energy is the energy set in motion by a vibrating body. For example, when a tuning fork is struck pressure waves that possess sound energy as they travel through the air are produced.

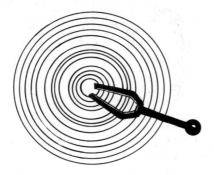

Heat energy is the energy an object possesses due to the motion of its molecules. As the amount of heat added to an object is increased, the molecules move faster and the object's temperature increases.

Light Energy is a form of energy often radiated by hot objects. The sun is our greatest source of light energy.

Transfer of Energy

Energy can change from one form to another. This changeability is an important property of energy. Water in the form of falling rain has kinetic energy. As falling rain collects in a pond, the kinetic energy of droplets forming the pond is changed to potential energy. If the pond is drained off by a stream or brook, these same droplets of water now moving downstream acquire kinetic energy. The kinetic energy of the traveling stream can be used to turn a paddle wheel of a wheat-grinding mill. The kinetic energy is thus transformed into mechanical energy. As the same water continues to move downstream, the kinetic energy of the moving water can be used to turn a turbine-generator which converts the kinetic energy to electrical energy. The process continues on and on. The electrical energy is changed by various devices in your home to heat, light, and/or sound energy.

The sun is responsible for most of our usable energy. Radiant and wind energy occur and recur, with minor interruptions, daily. These forms of energy are not capable of being stored unless converted to some other form of energy. The sun, however, also furnishes us energy indirectly. Energy from the sun is essential to the growth of plants, trees, and other living organisms, all of which form the major bases for the formation of coal, gas, and oil. These energy sources, however, are not daily recurring events — they take millions of years to form. We are consuming energy sources, which are not readily replaceable, at an alarming rate.

The earth receives approximately 250,000 calories per square centimeter per year from the sun. So much of this is sent back to outer space by reflected light or radiated heat that only 130,000 calories, or about fifty-six percent, reach the land or water surfaces. Of this 130,000 calories, 55,000 calories are returned to outer space as radiated heat and 5,000 calories are immediately reflected, leaving 70,000 usable calories per square centimeter per year. Since most of the 70,000 calories power the hydrologic cycle, the earth gives back to outer space, as heat and reflected light, about the same amount of energy as it receives from the sun.

Energy Terms

People communicate through their language. Areas of study such as medicine, law, science, etc., all have their own language. In order to communicate with these various groups you must learn their language. Energy has some terms peculiar to the study of energy. For example, a foot-pound is not a pound of feet. And, Btu is not another university with a good basketball team. In order to communicate about energy, you need to know a few simple terms.

Work is done whenever a force moves an object through a distance.

$$\text{Work} = \text{force} \times \text{distance}$$

Foot-pounds are units used in measuring the amount of work done.

Horsepower is a measure of work. One horse can move (raise) a 550-pound weight one foot in one second, or 33,000 foot-pounds per minute. Horsepower is the measure of work done by a horse in a given period of time. It thus is convenient to refer to this measure and use it as a reference point for comparison. For example, a car's engine can be measured in the number of horsepowers it is equivalent to.

Input is that which is put into something, for example, the power supplied to a machine.

Kinetic energy is the energy of motion.

Perpetual energy is the form of energy that is continuing or enduring forever.

Transfer of energy is the exchange of energy from one object to another. For example, when a hot object touches a cooler object, the faster moving molecules of the hot object lose or transfer energy to the slower moving molecules of the cooler object.

Power is the energy or force available to do work. This is the work accomplished or transferred per unit of time.

Watt is a unit of electric power equal to one joule per second. A watt has the power of a current of one ampere flowing across a potential difference of one volt.

Output is the quantity or amount produced and measured in a set time. For example, the assembly line at a major motorcar plant can turn out so many automobiles per hour or per day. This number of cars is referred to as the output or number for a specific unit of time.

Potential energy is stored energy.

Conservation law is a law that states, "energy cannot be created or destroyed, but can only be transformed from one form to another." This means that no new energy is created but rather that a fixed amount of energy exists and is transformed from one form to another with the initial amount of energy remaining the same. All forms of energy are interrelated and interconvertible. The conversion of one form of energy into another goes on continually. Energy is constantly released and stored in living systems. The success shown by some organisms over others is often due to the higher degree of efficiency with which they make use of the energy available to them.

British thermal unit One Btu is the amount of energy required to increase the temperature of one pound of water one degree Fahrenheit. The Btu of measure thus serves as a reference point from which to compare other energy producing materials. If you were to burn a one-pound piece of coal, you would release about 13,500 Btu's of heat energy. Similarly, burning a gallon of crude oil will yield about 14,000 Btu's and burning a cubic foot of natural gas gives about 1,000 Btu's.

Joule is a unit of measure. A joule is a one watt-second (approximately 0.74 foot-pounds).

Thermal gradient is a source of energy produced by the temperature differences between the deep and the surface of waters of the ocean, or the differences between the deep and the surface rocks of the earth. This is particularly important in tropical waters where high surface ocean temperatures exist.

Kilowatt hour One kWh is 1,000 watts of power used for one hour.

DID YOU KNOW?

Electrical energy is usually measured in watt-hours, but the Btu can also be used; one watt-hour is equivalent to 3.413 Btu's.

PEANUT BUTTER AND JELLY RACE

Work takes energy. No one (or few people) like to work longer and harder than they have to. Work takes energy and energy can sometimes be hard to come by.

The Task: To Increase Output

You and others are assigned the task of producing a large number of sandwiches for a school picnic. One student suggests that each person individually make as many sandwiches as quickly as possible. He feels that this is the quickest way to complete the task.

You argue that this is inefficient. You suggest an assembly line approach to preparing the peanut butter and jelly sandwiches. You challenge the other person to a comparison of the two proposed methods.

The Experiment

You set up two tables. At one table you allow eight members of the class to individually make sandwiches as fast as they can. You furnish them with all the materials: peanut butter, jelly, butter, knives, bread, and wrapping paper.

You select eight members to work assembling sandwiches at your table. You appoint each one to do a specific task; for example, one person butters the bread.

When both tables are fully organized and ready to start, and a specific time has been set, have your teacher say, "Ready, get set, go." Start sandwiching!

Which group turned out the most sandwiches?

What was your group's output per minute?

What was the opposition's output per minute?

BATHING

"To bathe or not to bathe" is not the question. Everyone should bathe. The question is whether you should shower or take a bath when you wish to clean yourself. Which method of cleaning uses less energy (and less water), a bath or a shower? Remember, it requires energy to get cold water to your home. Also, the energy necessary to raise this water to bath temperature is considerable.

What To Do

Fill your tub to the level that you feel you require to take a bath. Vertically insert into the water a meter (or yard) stick. Using a pencil, mark the level of the water on the stick. Next time you need to wash, take a shower. Make sure the drain is closed, so you can catch all the water you use while taking a shower. When you are done showering, step out of the tub and insert the same stick into the water. Mark the level of the water on the stick. How does it compare to the level of your previous tub bath? Which method of cleansing yourself uses more water and more energy?

What Would You Do?

What if the electricity and gas coming into your home were turned off daily from 7:00 P.M. to 7:00 A.M.? How would you adjust? How would you do the following activities differently?

prepare your food

store your food

wash your clothes

wash your dishes

heat your home

cool your home

entertain yourself (without TV, radio, hi-fi, etc.)

light up your home

What Would You Do?

What if the electricity and gas coming into your home were turned off permanently? Explain your response. How would your adjustments vary in the winter and summer months?

DID YOU KNOW?

If all the automobiles in the United States were run simultaneously, their electrical generating capacity would amount to fifty times that of all our central power stations.

DID YOU KNOW?

Most automobile engines are less than thirty percent efficient.

FACTS TO THINK ABOUT. WHAT'S GOOD FOR WHOM?*

A lot of wildlife has been killed in recent years because of man's scientific advances. Many wild animals have fled to land not yet touched by man. In Hidden Valley, Alaska, the last herd of Malibar Moose has found a safe home. Once abundant in the cold regions of North America, the Malibar Moose have been driven away or killed by advancing civilization. This is their last stand.

The Great Alabaster Oil Company has discovered an oil deposit in Hidden Valley and is taking steps to begin drilling.

Each person (or group of people) assumes one of the following roles or a role of his own choosing.

> Head conservationist for the State of Alaska
>
> Driller for the Great Alabaster Oil Company
>
> Governor of Alaska
>
> Person in charge of discovering new oil deposits for the Great Alabaster Oil Company
>
> Head of "Save Our Wildlife Society"
>
> Eskimo from near-by village
>
> Townsman from town located near Hidden Valley
>
> Secretary of Commerce, State of Alaska
>
> Anti-pollution committee member

Permit the students to raise questions and concerns relevant to the specific roles. Some suggested questions you *might* pose are:

> Do we have a problem?
>
> Who will benefit from the oil deposit?
>
> Who will not benefit from the oil deposit?
>
> What are some of the short- and long-term gains?
>
> What are some of the short- and long-term losses?
>
> How can we settle the problem?
>
> Do we need additional information?
>
> How many other questions can you raise (and answer)?

Energy Questions and Exercises

- Which sport — track, football, basketball, or baseball — do you think uses up the most energy?
- Does it take more energy to tell the truth or to tell a lie?
- How can energy be measured?
- Look around you, select any three objects, and explain how you could increase the energy of each object.

---

*Alfred De Vito, "What's Good For Whom?" *Science & Children*, vol. 10, #5, Jan./Feb., 1973, p. 32.

- How can energy be stored?
- Explain how the energy from the ear of corn you eat can be traced back to the sun.
- Trace and list the various forms of energy involved in processing an ear of corn into buttery, hot popcorn (from the sun's energy to finished product).
- Demonstrate as many different forms of energy as you can with a ball and a board.
- Which do you think uses more energy: laughing or crying?
- Identify and name all the forms of energy contained within yourself.

ENERGY EXPENSE ACCOUNT

ENERGY COST OF EXERCISE

Sport or Exercise	Total calories expended per minute of activity
Climbing	10.7–13.2
Cycling	
5.5 mph	4.5
9.4 mph	7.0
13.1 mph	11.1
Dancing	3.3–7.7
Football	8.9
Golf	5.0
Gymnastics	
Balancing	2.5
Abdominal exercises	3.0
Trunk bending	3.5
Arm swinging	6.5
Rowing	
51 strokes per minute	4.1
87 strokes per minute	7.0
97 strokes per minute	11.2
Running	
Short distance	13.3–16.6
Cross-country	10.6
Tennis	7.1
Skating (fast)	11.5
Skiing	
Moderate speed	10.8–15.9
Uphill, maximum speed	18.6
Squash	10.2
Swimming	
Breaststroke	11.0
Backstroke	11.5
Crawl (55 yards per minute)	14.0
Wrestling	4.2

Using the information in the preceding table record all the exercise you do in one day in the following chart.

Name of exercise	Time in minutes per day	calories per minute of activity	Total calories used per exercise
_____	_____	_____	_____
_____	_____	_____	_____
_____	_____	_____	_____
_____	_____	_____	_____
_____	_____	_____	_____
_____	_____	_____	_____
_____	_____	_____	_____

Grand Total: _____

HOW TO BUILD A RUBBER-BAND MOBILE

A rubber band does not create energy. It possesses energy, principally kinetic energy. Stretch it, and you put additional energy into the system. Its kinetic energy is increased. Therefore, its potential to do work is also increased. A stretched or twisted rubber band is capable of doing a surprising amount of work.

Materials

a thread spool
two wooden kitchen matches
a rubber band
a small candle for constructing a wax washer

Procedure

Cut the heads off the two wooden matches. Cut one match so that its length is shorter than the diameter of the spool. The longer match's length should exceed the diameter of the spool.

Put a rubber band, whose unstretched length is about twice that of the spool's length, through the hole in the spool. Place the smaller match stick through one of the looped ends of the rubber band.

Cut a small candle at right angles to its length, and fashion a wax washer with its diameter about one half that of the spool's diameter. Cut a small hole in the center of your candle washer.

Place the remaining loose end of the rubber band through the candle washer and put the longer match stick through this loop. The candle washer acts as a lubricant for the operation of the rubber-band mobile.

Rough up (notch, or roll the rims in glue and quickly run the rims through fine sand and let harden) the outer rims of the spool to improve traction.

Using the longer match stick as a lever (or crank) turn the match stick so that the rubber band becomes twisted many times. Record the number of revolutions you make. Place the rubber-band mobile on a flat surface and let it become mobile. How far per number of revolutions did it go? What happens when you vary the surfaces the rubber-band mobile must travel? The angle of steepness of a rise in elevation of various surfaces? The number of revolutions? The thickness of the rubber band?

How much work can your rubber-band mobile do? Design a technique to measure how much weight your rubber-band mobile can pull over various surfaces and inclines.

Trace the energy chain from the sun to the movement of the rubber-band mobile.

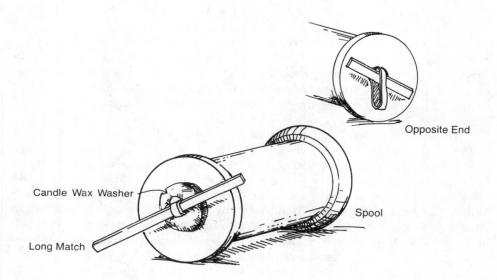

Opposite End

Candle Wax Washer

Spool

Long Match

FACTS TO THINK ABOUT
The silent guardians of Easter Island

History is punctuated with structures that baffle investigators such as anthropologists, geologists, historians, physicists, and curious observers who wonder how and why they were constructed. The magnitude of many of these structures boggles the mind. Ancient people, located in various places in the world, used excessive amounts of human energy to construct a variety of massive structures.

A peculiar drawing was carved high on the red wall cliffs south of Lima, Peru. Viewed from the sea, this 820-foot high carved drawing can be seen twelve miles away. The carved figure resembles a gigantic 3-pronged, fish spear.

Another interesting structure, whose construction has puzzled archeologists, is a series of circular settings of large standing stones called Stonehenge. Stonehenge is located on the Salisbury Plain in England. The innermost ring consists of five pairs of gigantic stones, each bridged with a horizontal piece. The vertical stones weighed approximately 40 tons apiece.

Another impressive engineering feat is the Egyptian tombs built in pyramid form. Stone blocks averaging 2½ tons each were used to build these pyramids. Some pyramids were made up of two million blocks. Occasionally, larger stone blocks were used—some as heavy as 15 tons apiece.

Not far from the sea, in the Peruvian spurs of the Andes Mountains, lies the ancient city of Nazca. A nearby valley contains a strip of level ground some thirty-seven miles long and one mile wide. If you fly over this territory you can make out gigantic lines laid out geometrically. Some of these lines run parallel to each other, others intersect or are surrounded by larger trapezoidal areas. Some archeologists say that these lines represent ancient Inca roads. If these are roads, they lead nowhere. They just end sharply. Some people say these lines represent some expression of religion or perhaps even a calendar. Other individuals have interpreted this construction as a landing field for ancient aircrafts that may have visited this planet from outer space. No one knows.

Perhaps the most interesting and challenging mystery of the past lies hidden on Easter Island.

Easter Island is the loneliest inhabited place in the world. It is located in the South Pacific Ocean about 2,000 miles west of the Chilean coast. It covers only 50 square miles and about 300 people live there.

Ancient people who once lived on Easter Island were called "long ears." They pierced their ears and dangled heavy weighted objects from them to stretch the lobes. Eventually their ear lobes reached down to their shoulders. Descendents of these ancient people still live on Easter Island. However, long ear-lobed people are not visible anywhere — only short ear-lobed people can be seen on the island.

Stranger than the disappearing long ear-lobes is the silent army of stone statues that seem to stand watch over the island, peering out to sea. There are over six hundred statues on the island. These giant statues are images of the long ears themselves. Every statue is exactly the same. All statues have the same facial expression. They are made of the same extremely tough, grayish-yellow black-grained stone. When this stone is struck with a steel axe, sparks fly. Every statue has a flat cut-off foundation. This cut-off is where the stomach ends and the legs begin. With one exception, all the statues are male figures. These statues are not found anywhere else in the world.

Each statue weighs an average of 50 tons (equal to about twenty-five to thirty medium-sized automobiles) and ranges in height from 30 to 69 feet tall.

The rock supply from which the giant statues were cut is the steep side of a water-filled, dead volcano. The volcano has been cut up like it was made of soft dough. Hundreds of thousands of tons of stone have been cut out and tens of thousands of tons of stone were carried away. This gaping hole contains more than one hundred and fifty gigantic stone giants in all stages of completion — some just begun and some just finished. The carved figures were cut in vertical and horizontal positions. It did not seem to matter to these ancient rock carvers in what position they worked. At the foot of the volcanic mountain finished stone men stand side by side, as if they had just come off an assembly line.

The major portion of the stonecutting and polishing of the statues was done at the volcano. The only stonecarving that was completed after having moved the statue to a desired spot on the island was the carving of the eyes. It was as though the stone carvers did not want the stone giants to see where they had been or where they were going. After being carved and polished, these statues (50 tons in weight and 40 feet tall) were moved all over the island. Some were moved as far as 10 miles away, over rough and hilly ground. Each statue was not only moved, but when it reached its destination, it was lifted up and placed on the top of a platform that was built 6 feet above ground level. The tremendous effort did not end there. A gigantic head decoration weighing from 2 to 10 tons (This is almost as much as the weight of two elephants. The average volume of the head decoration is approximately 200 cubic feet.) was placed on top of the silent stone giant's head. This decoration was similar to the hairdo worn by male natives of Easter Island. The color of the head decoration must have been important. The giant head decorations were cut from still another volcanic quarry which contained only red rock. This red rock quarry is located 7 miles from the supply of rock used for the body of the statues.

Facts

- It would appear that the ancient natives had no knowledge of the wheel, no cranes, no motorized vehicles of any kind, no metal tools, and no roads as we know them.
- It looks as though all the ongoing work stopped abruptly. Thousands of primitive, unpolished stone picks still lie out in the open. All stages of carving are visible. It doesn't look as if they were planning to stop the business of producing stone giants.
- Some head decorations were found in the shallow edges of the island's water.
- Bones of dead natives were found at the feet of these giant stone statues.

How was this all accomplished before the age of modern machinery? Is this possible without the use of machinery? These statues were not made by wood carvers who simply changed from carving wood to carving rock. These statues were cut from rock by people with experience in cutting and moving these giant statues.

- How did the people of Easter Island carve these huge statues out of this rough rock using only crude stone picks?
- How did they move the huge blocks of rock down the mountainside and then over land, sometimes miles?
- How could they get the enormous head decorations on top of the vertical statues?
- How do you think they polished these stone giants?
- Why did they polish these stone statues before moving them?
- Why did the ancient natives want long ear lobes?

When the present day natives on Easter Island are asked how they think the statues were transported, they simply state, "They went themselves." What do you think? Whose energy did what?

HOW MUCH ENERGY DOES IT TAKE TO BUILD A PYRAMID?

So you want to build a pyramid. Simple, just pile up some rocks and there you are. That is not the way the ancient Egyptians did it.

How do we think the Egyptians built the pyramids? The largest pyramid, the Great Pyramid of Cheops, was built by Cheops, the Pharaoh of the Third Dynasty. This pyramid is 755 feet along one side of its square base and 481 feet high. This structure would cover 13 acres of ground or approximately 64 square city blocks. The Great Pyramid contains about 2,300,000 blocks of limestone and granite weighing from 2 to 70 tons apiece and averaging 50 by 50 by 28 inches in size. These blocks were stacked stairlike in approximately two hundred steps to a height of a modern forty-story building.

Little is known of the entire process of pyramid building except that it took a lot of energy and work. The process of cutting the stone blocks at the quarry, the transportation of these blocks, and the erecting of these monuments was such a common, ordinary task for the Egyptians, they did not always consider it necessary to record how the job was accomplished. They simply went ahead and did it.

It is estimated that it took about twenty years to build a pyramid the size of the Great Pyramid. Using 2,300,000 blocks, how many blocks a year must be placed in position to complete the pyramid in twenty years? How many blocks must be positioned a day? In a twelve hour day, how many blocks per hour?

From an ancient tomb painting showing one hundred and seventy-two men moving a 60 ton stone statue, it is calculated that eight men could move an average 2½ ton pyramid block. Thus, it is thought that teams of eight men were organized to work on individual blocks of stone. Various teams of eight men cut the block at a stone quarry. Another eight men transported the block to the pyramid building area. And still another eight men were responsible for raising the block to the required position on the pyramid. Various teams of eight men could complete this process for ten blocks in five months. If eight men could add ten stone blocks in five months to the pyramid, how many men will it take to furnish the necessary stone blocks for one year's work on the pyramid?

How were the pyramids built? Many explanations of how the Egyptians moved the large, stone blocks have been proposed. Some say ancient priests of Egypt had the mental power to make the huge stones light and actually float them through space into position. Other people say the Egyptians had antigravity machines which made lifting huge stone blocks a simple task. Most people, however, think the job was accomplished with nothing but primitive tools, patience, and unlimited manpower.

Pulleys were unknown in ancient Egypt. In quarrying and building, workmen used copper chisels and possibly iron tools, as well as flint, quartz, and diorite rock pounders. The only additional aids were large wooden crowbars and, for transportation, wooden sleds and sometimes rollers.

To quarry the rock from the hillside, the Egyptians chipped away vertically with a wooden mallet and a copper chisel, which must have been highly tempered by some method unknown today. The Egyptians cut a groove in the rock where they wanted it split. They then drilled holes along the groove, beat wooden plugs very tightly into the holes and poured water over them. When moistened these wooden wedges expanded to crack the rock. Sometimes fires were built along the grooved lines, and water poured on the heated stone to obtain a clean break in the rock.

Even with adequate, though simple tools, patience, and unlimited manpower, the great organizing ability to start and finish a project such as a pyramid had to require a very special skill. It would take a genius' talent to plan all the work, to lay it out, to provide for emergencies and accidents, to see that the men in the quarries, on the boats, and sleds, and in the masons' and smithies' shops were all continuously and usefully employed, that the means of transportation was ample, that the food supply did not fail, that the water supply was adequate and conveniently available, and that the sick relief were on hand and available for replacement. It is estimated that as many as forty thousand skilled workers lived permanently on the spot. Feeding, housing, and clothing these workers was a tremendous undertaking.

There are many differing opinions about how the Great Pyramids were actually constructed. For the lack of equipment like steel cranes and derricks to lift and swing heavy blocks of stone, the Egyptians would have had to construct a ramp or ramps to raise heavy stones to the required level. Building the ramps was almost as great a task as building the pyramid itself. Some of the several ramp constructions suggested by Egyptian historians follow.

• As the pyramid rose, the builders raised an earthern mound on all sides of it, with one or more long ramps for hauling up the stones. As each row of blocks was laid, the mound and the ramps were raised another level. When the pyramid was completed, all this vast amount of earth had to be hauled away.

- By using the building itself as a ramp, the Egyptians dragged the stones up the pyramid's own spiraling outer wall. This would enable the builders to fill in the core as they went up and finish the casing as they came down.

- As the layers of the pyramid began to rise, four ramps made of stone rubble and mud were also built. Each ramp started at one of the four corners and rose to the unfinished level. The slope of the ramp was always maintained at 15 degrees.

Which ramping technique would you use? Having only simple machines such as the roller, the lever, and the inclined plane (the Egyptians made little use of the wheel), what techniques would you use to raise these large stone blocks into position? What happens to the length of the ramp as the elevation of the pyramid increases and the slope of the ramp is maintained at 15 degrees?

The limestone blocks and rubble for the interior of the pyramid appear to have been cut from local outcrops. This was dragged directly to the building area on sleds. To make the sledding easier, a liquid (probably milk, the fat content of which makes it a better lubricant than water) was poured on the ground in front of the path of the sled. Finer limestone had to be rafted from across the Nile River, almost 20 miles away. Granite used for the linings of chambers was floated on barges down the Nile from Aswan almost 500 miles south of the Great Pyramid.

For whatever reasons the Egyptians built pyramids, they expended a tremendous amount of energy to assemble such impressive monuments. The pyramids are monuments to energy as well as monuments to the ingenuity of the Egyptian people.

What Are the Sources of Energy?

Our way of life, as well as our lives, depend upon the sun. Nearly all the energy available to us is, or was, created by solar energy. The coal, oil, and gas that have fueled the expansion of our civilization were made from plants that acquired their energy by converting sunlight. Even the winds that drive our windmills originate from uneven heating of our atmosphere by the sun. Waterpower is dependent upon rainfall which is possible only because the sun's heat evaporates water standing on earth. The trees and other plants from which man and animals obtain energy, collect that energy from sunlight.

There are only three non-solar sources of energy: geothermal (heat from the earth's interior); tidal (the rise and fall of the tides due to the moon's gravitational effect); and nuclear.

DID YOU KNOW?

Did you know that there are four rules of ecology that must be kept in mind when considering energy choices:

1. Everything is connected to everything else.
2. Everything must go somewhere.
3. Nature knows best.
4. There is no such thing as a free lunch.

Each of the sources of energy has its problems. The trouble is, we don't have the luxury of that perennial alternative, "none of the above."

It has been estimated that the sun will last for another five billion years in its current state as a normal, or main sequence star. Thus, we might consider solar energy to be inexhaustible. Scientists generally agree that nuclear fusion is the source of energy of the sun and most other stars. Deep within the interior of the sun, temperatures are approximately 16 million degrees Celsius (25 million degrees Fahrenheit) and pressures are 70 trillion grams per square centimeter (a trillion pounds per square inch). Under these enormous pressures and temperatures, four hydrogen atoms are believed to fuse into one heavier atom of helium. The loss in total atomic weights between the original four hydrogen nuclei and the helium nucleus (about 0.7%) is transformed into an equivalent amount of radiant energy.

DID YOU KNOW?

Scientists theorize that about five billion years from now the sun will have depleted the hydrogen fuel in its core. Its thermonuclear reactions will then move outward where unused hydrogen exists. As it does, the tremendous nuclear heat at its core will also move outward, expanding the sun as much as sixty times. As the sun cools by expansion, its surface color will become a deep red. It will then be classified as a red giant star instead of a main sequence star. Looming across much of our sky, it will boil off our water and air and incinerate any remnants of life.

When the sun exhausts its hydrogen fuel, it will no longer be able to withstand gravitational contraction. Eventually it will shrink to a white (hot) dwarf, no bigger than earth, but so dense that a piece the size of a sugar cube would weigh thousands of kilograms. Eventually, after billions of more years, our sun will cool and dim to a black cinder. Only then will eternal night fall upon our solar system unless mankind creates some new energy source to replace our sun.

An Energy Time Line

Beginning with the Industrial Revolution and during the period of 1750–1850, the demand for large quantities of energy began, especially from England, western Europe, and America. Energy comes from many sources which are broadly summarized in the following list.

Type of Energy	Sources
Solar	The sun
Chemical	Petroleum, coal and wood
Motion	Running water, including waterfalls, tidal basins, man-made dams, and wind
Geothermal	Geysers and hot springs; hot water at great depth
Nuclear	Uranium, thorium, hydrogen, and others

In the first century B.C., man was able to generate the power necessary to light the equivalent of three 100-watt light bulbs (0.3 kilowatts) using a horizontal waterwheel. Setting the wheel vertically raised the generator's capacity to 2000 watts (2.0 kilowatts). When windmills were introduced, the power generated jumped to 12.0 kilowatts. Improvements in the waterwheel design raised the power to 56.0 kilowatts. This form of power generating continued up until about two hundred years ago. The use of power-driven machinery in England started man on a quest for more energy with which to run his machines and do his labor. In the middle 1700s, the steam engine, windmill, and waterwheel could generate an average of 500– 1,000 kilowatts of power. Our modern steam and gas turbines can generate up to 1,000,000 kilowatts of power.

DOES IT MAKE SENSE?

U.S. Energy Reserves and
Resources (Quadrillion Btu)

	Reserves	Resources
Coal	9,000	66,000
Oil	300	16,400
Natural gas	300	6,400
Shale oil	900	147,000
Uranium: thermal	200	440,600
Uranium: breeder	11,000	220,030,000
TOTAL	21,700	220,706,400

NOTE: Annual Consumption—77 in 1975, 163 in 2000

Reserves are materials which have been accurately located and proven to be economically recoverable using current technology.

Resources are much greater quantities of materials that include reserves as well as deposits. These deposits can be known but uneconomic, or unknown but inferred from geology or the existence of nearby reserves.

By the 1850s, wood was used to provide 90% of the energy used in the United States. The rapid depletion of forests and the need for better sources of energy caused a switch to coal energy which provided 70% of the energy consumed in the United States by 1900. By 1970, petroleum and natural gas supplied the United States with over 75% of the energy it consumed. By the year 2000, it has been estimated that 50% of the total energy consumption in the United States will be coal and nuclear energy.

DID YOU KNOW?

If every automatic dishwasher in the United States was run one less load a week, we would save the equivalent of about 9,000 barrels of oil a day—enough to heat 140,000 homes during the winter.

The fossil fuels (petroleum, natural gas, and coal) are so named because they are the result of a long-term distillation of plant and animal remains within the earth's crust. Natural gas and crude oil are the main components of petroleum. There are over two thousand petroleum products. Petroleum deposits can be found throughout the world and commercial production exists on nearly every continent. However, petroleum deposits occur in two large belts: one from Alaska to Canada, down through the Western United States and Gulf Coast, to Venezuela to Argentina; and the other from the Mediterranean through the Middle East to Indonesia. It has been estimated that the world's petroleum reserves will be depleted by the year 2000. Natural gas is usually found with petroleum deposits; however, it is possible to find one without the other. Natural gas is made up of several lightweight chemical compounds of hydrogen and carbon (hydrocarbons), the chief constituents being methane, propane, ethane, and butane. Natural gas burns readily in the presence of air and, in the process, the carbon and hydrogen molecules break up into individual atoms of carbon and hydrogen which then combine with the oxygen in the air to form carbon dioxide and water. The process of breaking up and recombining with oxygen is accompanied by the release of heat. This heat is used for our homes, cooking, and by industry. In fact, the biggest consumer of natural gas is industry (about 50%). By the year 2030 or 2040, our world's natural gas supplies may also be depleted.

DID YOU KNOW?

It takes a lot of materials and money to drill the typical 10,000 foot oil well. Some of the necessary materials are:

- 14,000 feet of steel pipe
- 11,500 feet of steel casing
- 20 drill bits
- 1,050 tons of drilling needs and additives
- 4,850 sacks of cement
- 48,000 barrels of water
- 3,000 barrels of diesel fuel

The average drilling cost per well rose from $88,554 in 1969 to over $150,000 in 1978. These costs do not include the cost of the drilling rig which is about $800,000.

Shale oil is petroleum derived from the destructive distillation of bituminous shales called oil shales. Petroleum can also be recovered from bituminous sands called tar sands. The difference between oil shale and tar sand is that the bituminous matter in oil shale is a solid whereas the bituminous matter in tar sand is a highly viscous liquid. In the United States, oil shale is found in the Green River Basin of Wyoming and in northwestern Colorado. The world's largest tar-sand deposits are found in a belt of sandstones extending over 960 kilometers (600 miles) in northern Alberta, Canada. The amount of energy that can be obtained from oil shale and tar sands may compare favorably with that from conventional oil and gas pools; however, recovery of petroleum from oil shale involves a number of environmental and economic factors not encountered in recovery of oil and gas through wells. Enormous quantities of rock must be removed and processed with some of the rock at the surface, while some is at depths of up to 600 meters (2,000 feet).

DID YOU KNOW?

Of the one hundred new field wildcat wells drilled in 1978, only nine found oil or natural gas, and only two had the one million-barrel reserve of oil or its gas equivalent required to make it commercially profitable.

DOES IT MAKE SENSE?

Does it make sense to burn oil to make electricity to send to a home to heat bath water?

Coal

Coal is one of the most important energy sources in the world and probably was the first fossil fuel to be used as a major source of energy. Early man undoubtedly used it on occasion. Coal is burned to release heat. One of its main uses is to provide heat for driving steam turbines which generate electric power. About half the coal used in the United States generates electric power, and much of the rest is used to make steel. It has been estimated that over 90% of the United States' energy reserves are in the form of coal and that there is enough coal left for about another one hundred and seventy to two hundred years before a peak in production will occur. Nearly all the coal will be mined approximately two hundred years after that peak.

In the United States, coal occurs in three regions: the Appalachian Mountains, the central region, and the Rocky Mountains. Coal occurs in stratified deposits in sedimentary basins with the individual coal beds (seams) usually being thin (no more than 3 meters or 10 feet) and laterally limited to a few kilometers (miles).

DOES IT MAKE SENSE?

A circus clown places a peanut on an overturned tub and has an elephant go over to crack the shell with his foot. The clown picks up the peanut and eats it. Everybody laughs.

In the energy world the peanut act is matched daily. At temperatures up to 1,200 degrees Fahrenheit we burn a limited resource to keep our livingrooms a comfortable 70 degrees. No one laughs when the bill arrives.

Coal is primarily composed of carbon, with smaller amounts of hydrogen and oxygen present. Some sulfur and nitrogen may also be found, however, sulfur is a highly undesirable impurity. Coal can be classified two ways: physically, or by rank (the heat value in the type of coal). The lowest coal rank is lignite; the highest is anthracite.

Coal forms from the progressive alteration of organic material that comes from plants that grew in swampy areas and died and decayed. As the decay began, masses of tree trunks, limbs, leaves, pollen, spores and other plant material piled up. Oxygen supplies were stifled by burial under water and under more debris. The wet mass of debris was turned into peat (a yellowish brown to black fibrous material in which fragments of plants are still recognizable). As the floor of the swamp continued to sink, debris piled up and the peat changed into brown coal which turned into lignite. The ultimate stage of development from lignite is high-rank bituminous coal, but if the beds were squeezed together, the lignite becomes anthracite.

DID YOU KNOW?

Solar, geothermal, and synthetic fuels will make only a small contribution to domestic energy supplies by 1985—about 1%.

Unless commercial-size solar, geothermal and synthetic processing plants are started now and proven economical by 1985, it will not be possible for these new energy sources to replace dwindling supplies of oil and gas in the post-1985 period.

DOES IT MAKE SENSE?

**COMPARISON OF TONNAGE OF CELLULOSE
WASTES WITH PETROLEUM CRUDE OIL**

Feedstock	Availability in U.S.A.
Cellulosic wastes	1.0 billion tons per year, renewable
Petroleum crude oil	0.7 billion tons per year, non-renewable

Nuclear Fission and Fusion

Since fossil fuels won't be able to meet the projected demand for energy in the future, we must either turn to alternative sources of energy or put a limit upon the amount of energy we use. From the first self-sustaining chain reaction and controlled release of nuclear energy in 1942, mankind became aware of the vast store of energy contained in the nucleus of the atom. In 1954, the United States Congress permitted the use of nuclear energy for the production of electricity. In 1970, about 1% of the United States' electrical power was obtained from nuclear energy. By 1976 nuclear energy yielded 9.4% of our electrical power. It has been estimated that by the year 2000 nuclear energy will supply 30% of our total energy supply, coal will provide 30%, oil and natural gas only 30% (from 75% in 1976). Solar, geothermal, oil shale, use of waste materials, and other energy sources will account for the other 10% in the year 2000.

DID YOU KNOW?

If there were one hundred nuclear plants in operation in the United States, the chance of an accident causing one hundred or more immediate deaths would be one in one hundred thousand. This figure does not count subsequent deaths from radiation, which would depend on the accident's site.

Man-caused and natural disasters capable of killing 100 or more persons are:

Man-Caused	*Natural*
Fire: one in two years	Tornado: one in five years
Explosion: one in seven years	Hurricane: one in five years
Toxic Gas: one in one hundred years	Earthquake: one in twenty years

Nuclear energy can be generated in two fundamentally different ways: fusion or fission. In fusion, energy is released in the process of fusing the nuclei of two relatively lightweight atoms to form a heavier atom. For example, various methods can be used to fuse isotopes of hydrogen to form helium. Natural hydrogen consists of three isotopes: protium, H^1; deuterium, H^2; and tritium, H^3. When tritium (H^3) and deuterium (H^2) are fused to produce helium (He^4), tremendous amounts of energy are produced. This process is the same as the sun's burning, and is an uncontrolled reaction. While controlled fusion has not been achieved yet, research is being conducted to achieve this goal.

When controlled fusion is accomplished, the problem of an energy source may be solved because one out of every five thousand atoms of hydrogen is deuterium. The sea will then become an inexhaustible fuel source. We will return to the fascinating topic of fusion in the section on alternative energy sources.

DID YOU KNOW?

The ocean is an almost inexhaustible fuel source for a fusion nuclear reactor. Since approximately one out of every five thousand hydrogen atoms is deuterium, the amount of energy obtainable from 1 cubic yard of sea water is equivalent to that from nearly one thousand barrels of oil. The energy of the deuterium in a cubic mile of sea water is of the same order of magnitude as is the energy of the world's entire petroleum supply.

In fission, energy is given off when heavier atoms are "split" into lighter atoms. One example is the splitting or fission of one isotope of uranium, $U-235$. If the $U-235$ nucleus is bombarded with a neutron, the atom may split into barium-141, krypton-92, neutrons, and energy. Because the neutrons are fast-moving particles, they penetrate the nuclei of adjacent uranium atoms and cause the reaction to continue in a similar manner; this process is called a chain reaction. Chain reactions will occur only if sufficient uranium is present; we refer to this amount of uranium as the critical mass.

The production of energy through fission can be accomplished in any one of three basic atomic reactor designs: burners, converters, and breeders. In a burner reactor, all the fissionable material, usually $U-235$ and plutonium-239, is consumed by the end of one fuel-consumption cycle. In a converter reactor, uranium-238 or thorium-232, neither of which is readily fissionable, is fed into the reactor along with the fissionable material contained inside this quantity of the uranium-238 and thorium-232. As the readily fissionable uranium-235 and plutonium-239 decay, they emit high speed neutrons. These neutrons bombard the uranium-238 and the thorium-232 which are then transmuted into plutonium-239 and uranium-233 respectively. Both of these end products are readily fissionable and usable as fuel.

If the fissionable isotopes are produced in amounts less than the original quantity of $U-235$ fuel, the reactor is called a converter; but if the amount is greater, it is called a breeder reactor. By fueling one or two breeder reactors with natural $U-235$, sufficient fissionable isotopes can be produced to fuel other breeder reactors. However, no commercial breeder reactors exist at this time. In 1976, we had about sixty-five nuclear power plants operating in the United States; by 1985 it may be as high as one hundred and seventy-five, and by the year 2000 as many as four hundred nuclear power plants operating. In addition, thirty-eight countries will be generating between 13% and 85% of their electricity by nuclear power in the year 2000.

Uranium occurs in nature as three principal isotopes, $U-234$, $U-235$, and $U-238$. The percentage of each isotope of uranium in nature is: $U-238$ (99.238%), $U-235$ (0.711%), and $U-234$ (0.006%). It is obvious from these percentages that burner reactors, capable of using only $U-235$, would soon deplete our supply of this isotope unless it becomes available in vast quantities. However, burner type reactors are the only type of reactors commercially used today and almost all of those in use or on order are light-water reactors (LWR) or those using ordinary H_2O in their internal system. Light-water reactors are of two types: boiling-water reactors (BWR) and pressurized-water reactors (PWR). Approximately 60% of the current nuclear power generation market uses pressurized-water reactors, while the remaining 40% uses boiling-water reactors.

Uranium ores occur in many countries including: the United States, Canada, Zaire (Africa), and Czechoslovakia. In the United States, the Colorado Plateau has yielded the majority of uranium produced in this country. Uranium ore is processed so that the uranium becomes a concentrate of the oxide mixture, U_3O_8. The amount of U_3O_8 is limited, therefore lower grade ores will have to be used in the future. The use of fission reactors involves the additional problem of what to do with the radioactive waste materials, some of which present difficult disposal problems.

DID YOU KNOW?

That a fusion reactor will present some difficult engineering problems? The fusion process requires a strong magnetic field and superconducting magnets that are maintained at very low temperatures. Therefore a device will have to be created which can have temperatures as high as 100,000,000 degrees Celsius at the center, and almost as low as absolute zero only two meters away.

We shall explore this and other alternative sources of energy such as gasification, geothermal, solar, tidal, hydrogen, fuel cells, gas from organic wastes, wind, and photosynthesis in a later section of this book.

In conclusion, it appears obvious that fossil fuels will become rapidly exhausted if they continue to be used as our major energy source. It is clear that we need to develop alternative sources of energy as soon as possible and it seems that a major source will be nuclear power. Nuclear energy is the only suitable alternative that is technologically capable of replacing our dependence on fossil fuels. However, without breeder reactors or a feasible fusion process, nuclear energy has its limits and could become a short-term alternative. In the coming decades solar energy will be developed as a usable resource along with several other alternatives. One can only speculate as to their form. Rather than witnessing an end of energy usage on our planet, we are seeing the beginning of a transition from the use of fossil fuels as our prime energy source to the use of some other resource such as solar energy. This transition will at times be painful, very expensive, and require a comprehensive education program in our communities so that future generations will be prepared to participate in this transition.

SOURCES OF ENERGY: WHAT'S YOUR LINE?
MAKE YOUR OWN ENERGY TIME LINE

Materials needed

    adding machine tape roll (or similar paper roll)

    colored pencils, crayons, or markers

    meter or yard stick

Procedure

Using a measured length of tape (1 – 2 meters), divide the tape into sections that illustrate the development of energy. For example, if 1 decimeter equals five hundred years, you should be able to mark off the energy highlights on your paper tape. Here are some that you might want to include:

6000–4000 B.C.	Prehistoric	Muscle energy
4000 B.C.	Ancient Egypt	Water/wind power
100 B.C.	Ancient Greece	Early solar power
		Improved muscle power
1200–1500 A.D.	Crusades	Wind/water power
1700–1850 A.D.	Age of Enlightenment	Steam power
1850–1950 A.D.	Age of Technology	Electrical power
1950–? A.D.	Our energy future ???	Nuclear power
		Solar power
		Tidal power

What will your generation's contribution to the energy time line look like? You may want to illustrate your energy time line by including representative energy sources for each of the major divisions mentioned.

More Time Lines

What other time lines could be developed? Try time lines illustrating energy-saving devices, alternative energy sources, conservation techniques, energy production (for example, the time it takes for an oil well to become a producing one, or the time it takes to construct a nuclear reactor power station).

WATER WHEEL POWER: TURN, TURN, TURN

Materials

water wheel (can be either purchased at a store that sells aquarium supplies or constructed)
large pan or dish to put the water wheel in (some schools have stream tables that could be used)
buckets
hoses
sand
clamps

Procedure

Place a 4– 6 centimeter layer of sand in the pan or stream table, leaving the drain end free of sand. Set up your pan or stream table as indicated in the following diagram.

Place the water wheel in a horizontal position in a steam bed channeled in the stream table. Using the fall of water as a source of energy, design a mechanism to get the water wheel to revolve. Variables that might be tested include: rate of water flow, position of the supply hose in relation to the water wheel, and the distance of the supply hose from the water wheel.

Now reposition the water wheel so that it is in a vertical position. Compare the beneficial effects achieved using this method. Investigate the variables previously mentioned when considering the horizontal water wheel. Graph the variable chosen (x– axis) versus the number of revolutions made by the water wheel (y– axis). It should be very evident that man's movement of the water

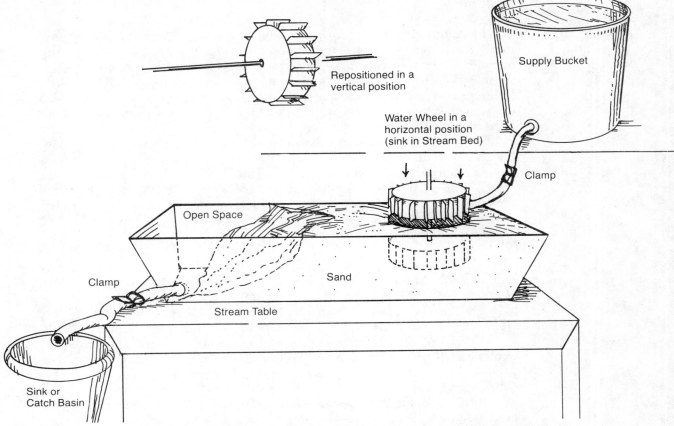

Repositioned in a vertical position

Supply Bucket

Water Wheel in a horizontal position (sink in Stream Bed)

Clamp

Open Space

Clamp

Sand

Stream Table

Sink or Catch Basin

wheel from a horizontal to a verticle position resulted in a significant increase in power available (number of revolutions) to him.

More

Have the students hook up their water wheel to something that will allow them to do work. For example, a string could be attached to the water wheel's axle with a sinker fastened to the other end. How many revolutions of the water wheel does it take to move the sinker a measured distance such as 10 centimeters? How many sinkers can the water wheel lift? What variables will affect this? How would a drought or low stream level affect the ability of the water wheel to generate power? Could the water wheel serve as a generator of electricity? What else could the water wheel be used for? Should we return to use of the water wheel as our primary source of electricity? What other advantages and disadvantages do you perceive?

STEAM POWER: I'M ALL STEAMED UP!

Materials needed

1 toothpowder or talcum powder metal container
1 small candle
1 plastic soapdish
wire (from a coat hanger)
1 large pan or stream table

Procedure

Place the candle in the plastic soapdish and make sure that it is secure. Punch a hole near the edge of the bottom of the metal container. Make sure this opening is always kept open. Construct wire legs that are long enough to support the metal container horizontally over the candle and the soapdish. Position the metal container horizontally, so that the needle hole is located at the top. Half fill the metal container with water, replace the cover, place this "boiler" over the candle and light the candle.

Place the entire apparatus in a pan of water and observe what happens. Can you explain the operation of this boat in terms of the principle of energy conservation? How does the amount of water affect the distance that the boat travels? Can the relationship between the two be graphed? How does the size of the container (or candle) affect the distance and speed that the boat travels?

More Facts About Steam Power

The steam boat that you made is patterned after the *aeolipile* invented by Heron of Alexandria in 100 A.D. It was a toy meant to entertain people. What else can you find out about Heron's aeolipile?

The first successful steam engine was developed by Tomas Savery (1650–1715). What can you find out about how the Savery engine worked?

The use of high pressure steam produces a serious risk of boiler or cylinder explosions. Thomas Newcomen (1633–1729) developed an engine that uses steam at lower pressure. Can you prepare a chart comparing the Savery engine with the Newcomen?

In 1764, James Watt was asked to repair a model of Newcomen's engine and ended up developing an engine that could do more than twice as much work as Newcomen's with the same amount of fuel. This improvement enabled Watt to make a fortune by selling or renting his engine. Watt's fee depended upon the power of the engine. Watt's steam engine helped transform the economic and social structure of Western civilization.

Why are steam engines no longer used as direct sources of power in industry or transportation? Steam is still the major source of power when you consider that steam turbines drive the electric generators in most electric power

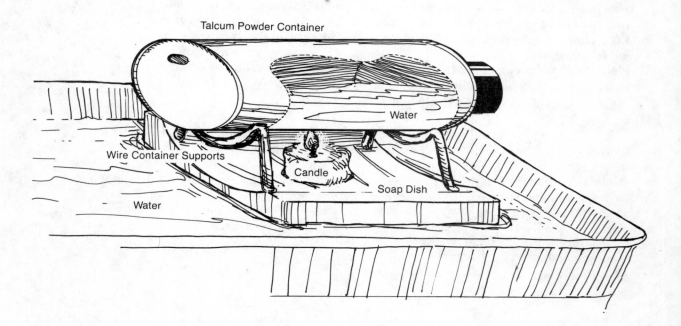

Talcum Powder Container

Water

Wire Container Supports

Candle

Soap Dish

Water

stations. Even in nuclear power stations, the nuclear energy is generally used to produce steam, which then drives turbines and electric generators.

Prepare a time line of the improvements made in the steam engine since Heron's time. Don't forget to include the contributions of Parsons and Joule.

THE DISTILLATION OF WOOD

(**Caution:** This experiment is only to be done under the direct supervision of the teacher)

Materials needed

> eye protection goggles
> 2 test tubes (Pyrex)
> 1 beaker
> 1 bunsen burner or alcohol burner
> rubber tubing
> glass bottle
> glass tubing (see Figures 1 and 2)
> wood splints
> boiling chips or broken porcelain

Wood is a familiar substance, and the burning of wood is a common occurrence when camping or using a fireplace. Wood was one of the first fuels used as a source of heat energy and yet very few of us have really observed wood closely. It takes some time to start a wood fire because the wood must reach a high temperature before it bursts into flame. You may have wondered how hot wood must get before it begins to burn. Although we usually burn wood in an abundant air supply, we might also ask: "What happens when we heat wood in a limited supply of air?" What you find out will depend on the design of your experiments and how carefully you observe.

Procedure

To observe what happens when you heat some wood splints in a closed test tube, set up your apparatus as shown in Figure 1. Before you begin, infer an outcome for the experiment. Pack a Pyrex test tube with wood splints and connect it to the apparatus by clamping each part to ringstands. Heat the tube with either a Bunsen burner or two alcohol burners (remember to wear protective eyeware). What do you observe happening when the wood gets hot? Will the gas that comes out of the tube burn? Use a glowing wood splint to test your answer.

Now attach a piece of rubber tubing to the apparatus as shown in Figure 2 and collect the gas in a bottle by the displacement of water. When the bottle is nearly full of gas, disconnect the rubber tubing. Relight the gas coming from the test tube and keep it burning as long as you can. Since the gas may come out in puffs, you may have to relight it frequently. You probably will have to tilt or move the burners to heat all of the wood. Stop heating the splints when you can no longer keep the gas lit.

Observe the gas that you collected in the bottle. How does its volume compare to that of the splints you used? What is its color? Is it soluble in water? If you wanted to determine its composition, you would need to perform several tests. Turn the bottle upright and set it in a well-ventilated place for a few minutes to get rid of the gas, and then pour out any residual water. *Avoid deliberately breathing the gas since it is poisonous.*

Examine the liquid in the upright test tube. Is it one liquid or more than one liquid? Disconnect the tube containing the condensed liquid and put a few pieces of broken porcelain in it (boiling chips). The porcelain chips help to keep the liquid boiling evenly. Connect the apparatus as shown in Figure 3 and boil off about half the liquid. What do you observe happens? Is the liquid that condensed in the right-hand test tube the same as that in the left-hand one? How do the two liquids compare? What do you observe happens when you mix them together?

After the test tube containing the wood has cooled, examine the remains of the wood splints. Describe the appearance of the wood splints. Take one splint out and hold it in a flame (BE CAREFUL). Does it burn? Does it burst into flame? Does it leave any ash? Does it resemble any substance familiar to you?

Refer back to your inference at the beginning of this activity. Did you infer that all of these gases and liquids would be obtained from the wood? Can you get the wood back by mixing all the substances you have collected? Were these substances actually present in the wood or were they formed by heating? Can you cite any evidence for your answer?

FIGURE 1 APPARATUS FOR DISTILLING WOOD

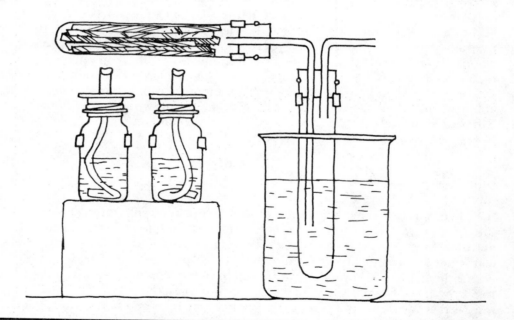

More Questions About Wood

- What are the advantages and disadvantages of using wood as our primary source of fuel?
- Why don't all homes have fireplaces like they did in the 1700s?

It has been stated that wood is the primary raw material in the development of civilizations and remains indispensable for many basic industries. Obtain evidence to defend or refute this statement.

FIGURE 2 APPARATUS FOR COLLECTING GAS

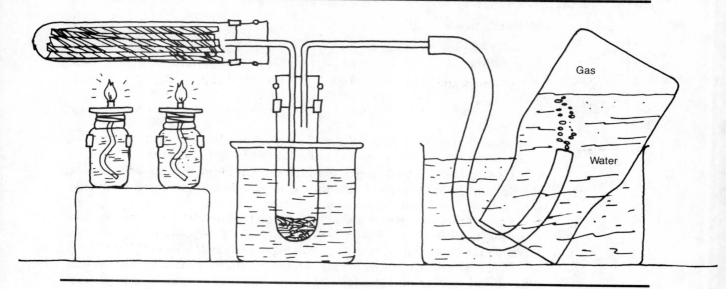

FIGURE 3 APPARATUS FOR REDISTILLATION

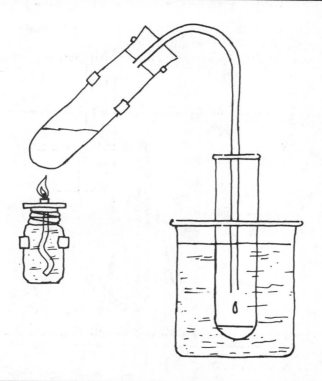

A chemical analysis of wood indicates that it is 50% carbon, 6% hydrogen and 44% oxygen. Identify these materials in your distillation of wood experiment.

- What is wood used for? How many different uses can you think of?
- Would you prefer to live in a home built entirely of wood or stone? Why or why not?

BUILD YOUR OWN OIL WELL

Materials needed

shoebox	soil
toothpicks	rocks
plastic wrap	oil (motor)
straws	

Procedure

Construct an oil well similar to that shown in Figure 4. You will note that the derrick is constructed of toothpicks, the drill and casing are made from straws, and plastic wrap is used to separate the layers.

Cut off the front end of a shoebox. Make a sketch of the subsurface (see Figure 4) to fit the back portion of your shoebox. Glue this in place. Replace the shoebox cover—it will serve as the base for the oil derrick.

What hazards can be expected when drilling for oil and/or natural gas? After the oil and/or natural gas leaves the well how are they transported to you?

More Facts About Oil

For every four barrels of oil we take from our own reserves in the United States, we import one barrel from a foreign country. Prepare a map which indicates the size, quantity and quality of the oil-producing areas of the world. Next, make a coal areas map.

FIGURE 4

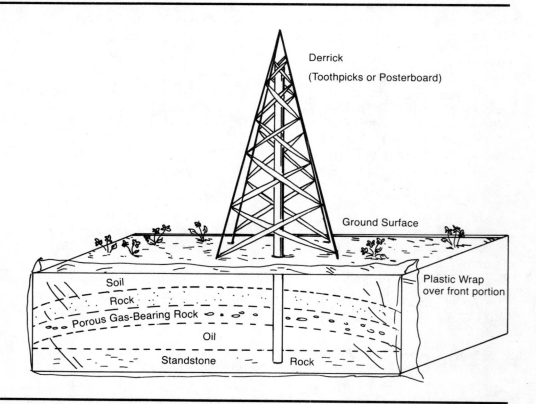

Recently beaches around the world have been blackened by oil washing onto their sandy shores from super-tanker accidents. Fish and wildlife have been killed by oil slicks. Place several drops of motor oil on a water surface. Observe what happens in relation to the sinking and floating of each. Once oil has spilled into a body of water, it is very difficult and expensive to remove. Devise a practical method for removing an oil slick from the water. Try it to determine if it works.

COAL—OR, I WAS BURIED 300 MILLION YEARS AGO!

Materials needed

aquarium	plant material (leaves, stems, roots)
coal	soil
peat	

The energy obtained by burning coal came from the sun millions of years ago. The energy in coal is nothing but the energy of sunshine, captured and buried in the ground long ago. To see how sunshine came to be stored in coal, we have to look back about 300 million years to the time when most of the coal was made. At that time the earth was warmer and more moist than it is now, and lush giant ferns and palm trees grew in vast forests and swamps where dinosaurs crashed through the woods. As these plants died and decayed, a thick layer of plant material was gradually built up and as time went on, a layer of peat was formed from the rotting leaves, stems, and roots. This layer was later covered by sand from a river, or by windblown dust until it was buried deep in the ground. Under the pressure of the soil above it, the peat gradually changed into coal. Still locked in it was the chemical energy that plants had stored. Today the original energy from the sun, stored so long ago, can be released when coal is burned.

Procedure

Place a layer of coal in the bottom of the aquarium for the first layer, then put a layer of peat on top of the coal. Next, place the plant material on top of the peat and cover it with soil. Now that you have constructed a model of how coal is formed, devise a way to get the coal out without disturbing the plant material, peat, or soil. See Figure 5.

An easy way to get this buried treasure is through the simple process of stripping off the layers to reach the coal. Put each layer in a separate pile so they won't mix together. This is called *strip mining or surface mining.* Strip mining requires the use of some of the most oversized equipment in existence. A huge shovel or dragline digs a long trench to get to the coal seam. Smaller shovels dig out the coal and load it into huge trucks for delivery to either a nearby "mine mouth" generating plant or a railroad loading. The largest of these shovels, "Big Muskie," can dig 325 tons of soil in one bite. Trucks or earth movers capable of carrying 100 tons are used. Now that you have taken the coal layer out, place the other three layers back the way they were, to simulate restoration of the land.

More Facts and Activities About Coal

There are three other types of mines: shaft mines, drift mines, and slope mines (see Figure 6). These are different kinds of underground mines. In most

FIGURE 5

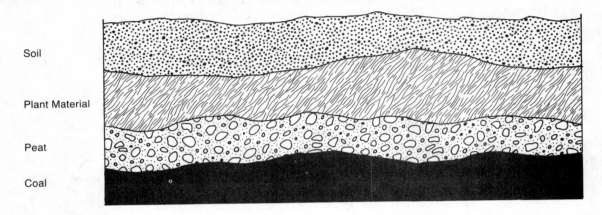

Soil

Plant Material

Peat

Coal

underground mines the undercutting is done by electrically-powered cutting machines that look like giant chain saws. After the undercutting, the coal is loosened by blasting. Loading machines scoop up the loosened coal and dump it into a moving belt, or into a waiting shuttle car which starts it on its trip out of the mine. In a shaft mine, the moving belts or shuttle cars carry the coal to the base of the shaft. Elevators then lift the coal out of the mine. In a drift mine, the moving belts or shuttle cars carry the coal straight out to the mine entrance in the side of the hill. In a slope mine, the coal is carried out of the mine by moving belts or electric railways that travel from the coal seams, along the slope tunnel, to the surface. Using Figure 6, select one type of coal mine and write a story about your imaginary visit to it. What do you think you would see? How would you feel inside a coal mine?

The following chart shows four different types of coal. Obtain some samples of coal, then see if you can identify them by using the chart. Tape the coal in the area provided on the right side of the chart. Classify each sample into one of the four categories. Describe the differences between them.

FOUR TYPES OF COAL

Kind of Coal	Physical Appearance	Characteristics	Sample
Lignite	Brown to brownish black	Poorly to moderately consolidated; weathers rapidly; plant residues apparent	
Sub-bituminous	Black; dull or waxy luster	Weathers easily; plant residues faintly shown	
Bituminous	Black; dense; brittle	Does not weather easily; plant visible with microscope; burns with short blue flame	
Anthracite	Black; hard; usually glassy luster	Very hard and brittle; burns with almost no smoke	

FIGURE 6 FOUR TYPES OF BITUMINOUS COAL MINES

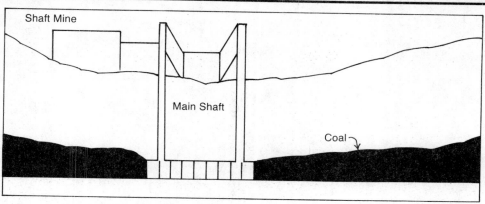

Shaft Mine

Main Shaft

Coal

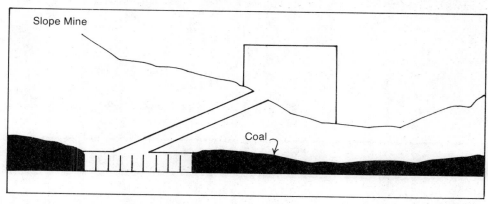

Slope Mine

Coal

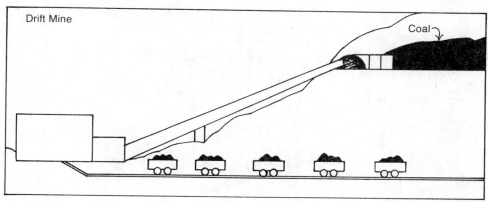

Drift Mine

Coal

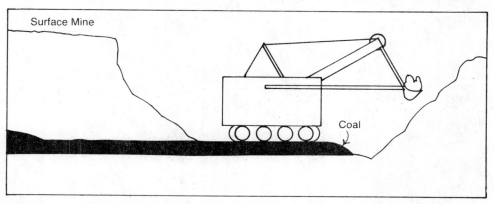

Surface Mine

Coal

RECOVERABLE COAL-RESERVES OF THE WORLD

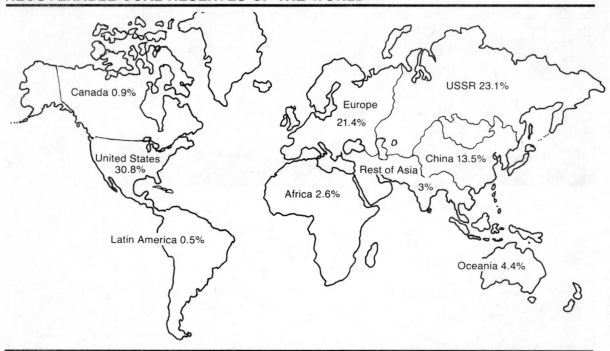

RECOVERABLE COAL-RESERVES IN THE UNITED STATES

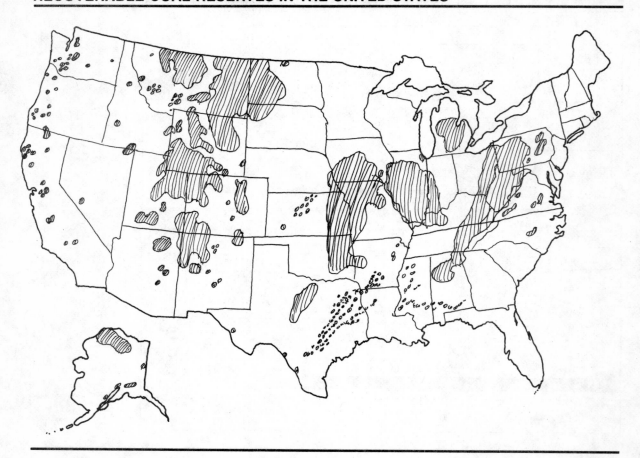

WHERE CAN WE FIND COAL?

Look at the recoverable coal-reserves map of the world and answer the following questions:

- What country contains most of the coal reserves?
- What country contains the least?

Look at the recoverable coal reserves map of the United States and answer the following questions:

- Where can coal be found in the United States?
- Obtain a map of your state. Does your state have any coal reserves?
- Where are they located? How will the coal be mined and transported in your state?
- What are the advantages and disadvantages of these processes?
- Which state holds most of our coal reserves?
- Why do you think the coal deposits are in these areas?

The reason these places have the most coal deposits can be traced back 300 million years. Coal was formed when the lush giant ferns and palm trees which grew in vast forests and swamps decayed and were buried. These places probably had a warm climate. As a result, more coal was formed in these areas than in the areas without forests, swamps and warm climates.

DIFFERENT TYPES OF COAL MINED IN THE UNITED STATES

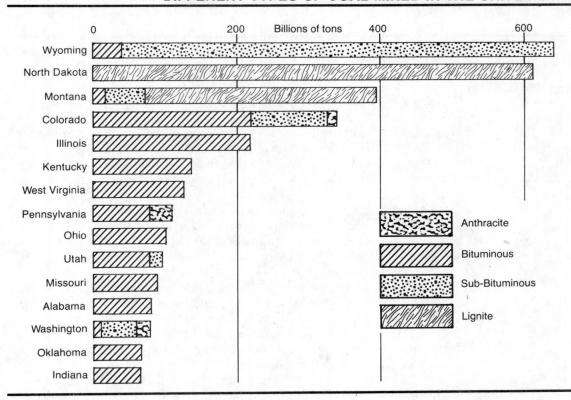

- Which type of coal is mined in your state?
- Which type is mined the most in the United States?
- Which type is mined the least?

THE PERCENT OF COAL USED AS FUEL IN AREAS OF THE UNITED STATES

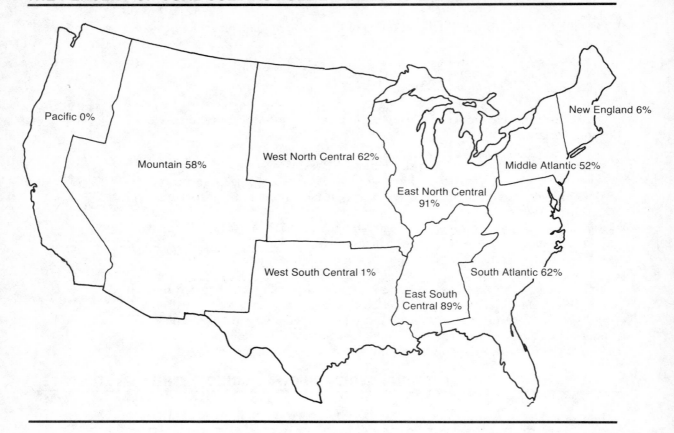

Pacific 0%

Mountain 58%

West North Central 62%

East North Central 91%

New England 6%

Middle Atlantic 52%

West South Central 1%

East South Central 89%

South Atlantic 62%

USES OF COAL

- Which areas use coal as fuel the most?
- Which areas use coal as fuel the least?
- Why do you suppose these areas do not use as much coal as the Midwest?
- The Pacific, West South Central, and New England areas use other resources, such as oil and gas, for their chief fuel because of their location. What other types of energy might they use?

Each of us uses about 12 pounds of coal every day. But most of us never see any of the coal we use. We use coal to make electricity. Consider the many different ways we use electricity. We use coal to make steel. Think of different items made of steel. We use coal to make cement. What do we make with cement? We use coal to make such products as rubber and man-made fibers, drugs and perfumes, food flavoring and dyes, plastics and waterproofing materials. In fact, more than two hundred thousand different products are made from coal.

Figure 7 is a tree showing the different uses of coal. Complete the chart by writing in some of the extended uses of coal. For example, coal branches into steel which is used to make automobiles.

Today 20% of our energy sources come from coal. Coal is America's most abundant fuel resource—about 80% of the nation's proved energy reserve—and the coal industry has been called upon to expand its production by at least a factor of 2 by the year 1985. Since it is so abundant, why haven't we relied more on coal for energy?

FIGURE 7 USES OF COAL

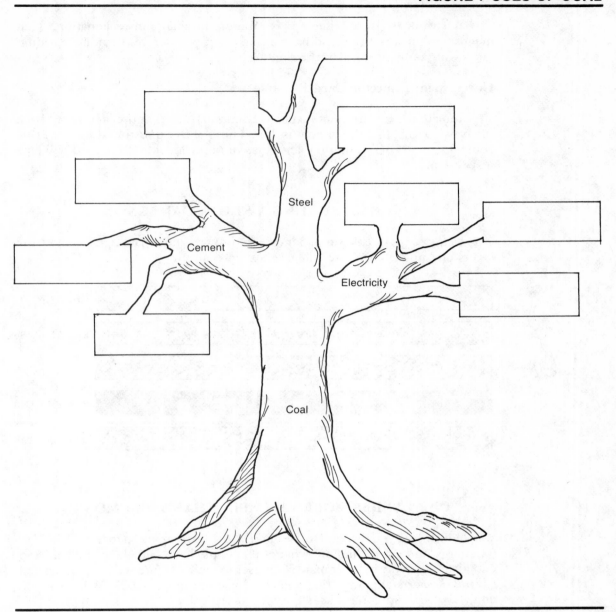

Steel

Cement

Electricity

Coal

COAL MINING ROLE-PLAYING

Use interlocking construction blocks to make the equipment used to mine and transport coal, such as the dragline, trucks, trains, loaders, bulldozers, etc. Then role-play the different people involved in the mining of coal.

Coal is a resource. Make a list of all the other resources we use or lose to get to coal and to get the energy from it. Remember to consider the land that will be changed; the machines to be loaded; methods of transporting coal; and the pollution from burning coal.

Divide into groups representing company personnel and government inspectors. Find dilemmas that coal companies and inspectors face and try to find a solution to each one. You may want to write the dilemmas on cards and hand them to each other to solve. Some examples of these cards follow.

Company Card Examples

You work for the coal company. You are in charge of reclamation. That means you must put the land back exactly as it was before the mine was dug. How will you need to plan the digging?

Government Inspector Card Example

You are the government inspector. You must see that the land is put back the way it was before the mine was dug. Think of five questions to ask the coal company team. Based on their answers, you must decide if the land will pass inspection.

CONSTRUCTING A STRIP COAL MINE

Using strips of colored paper (green, brown, gray, black, gray), show the layers one might encounter digging a strip coal mine.

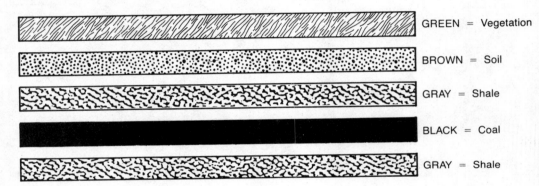

GREEN = Vegetation

BROWN = Soil

GRAY = Shale

BLACK = Coal

GRAY = Shale

OUR LAND IS MORE VALUABLE THAN YOUR MONEY

How would you respond to the chief's feelings in relation to the use of land for energy purposes? Find out what other people have said about the land and energy usage in the past. Prepare responses to their statements. Prepare several *Did You Know?* statements that you could use in the year 2025. Will what you don't say be as important as what you do say?

DID YOU KNOW?

"Our land is more valuable than your money. It will last forever, it will not even perish by the flames of fire. As long as the sun shines and the waters flow, this land will be here to give life to man and animals. We cannot sell the lives of men and animals; therefore we cannot sell this land. It was put here by the Great Spirit and we cannot sell it because it does not belong to us. You can count your money and burn it within the nod of a buffalo's head, but only the Great Spirit can count the grains of grass on these plains. As a present to you, we will give you anything we have that you can take with you; but the land, never."

Blackfeet Chief, Recorded in a
19th Century Treaty Council

I'M GOING NUCLEAR OVER YOU: CONSTRUCTING A FISSION NUCLEAR REACTOR

Materials needed

 shoebox

 4 empty toilet paper cardboard tubes (or 2 empty paper towel cardboard tubes cut in half)

 2 tongue depressors or popsicle sticks

 2 soda straws

 crayons

 glue

 scissors

Procedure

 Color two of the toilet paper cardboard tubes. Do not color the remaining two tubes. Color the inside of the shoebox blue. Glue or tape the tubes to the bottom of the shoebox (see Figure 8). Leave enough room for another tube to fit between them. Glue a tongue depressor or popsicle stick to the inside of each of the uncolored tubes. Make sure that most of the popsicle stick is sticking out from the tubes. The two uncolored rolls with the popsicle sticks must be placed in between the other two rolls as shown in Figure 8. The popsicle sticks should be sticking out of the top of the box so that you can move these two rolls up and down by pushing or pulling on the sticks. On each end of the box, make a small hole with your scissors and put a soda straw in each of these holes. Your nuclear reactor is now complete! Let's find out how it works.

FIGURE 8 FISSION NUCLEAR REACTOR

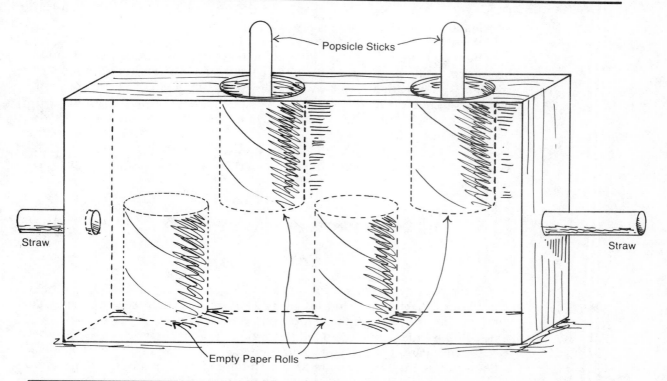

In your nuclear reactor, the colored tubes are called the fuel rods. In these fuel rods, neutrons are splitting atoms, causing more neutrons to be released. These collisions generate heat.

The movable uncolored tubes are the control rods. When the control rods are all the way in, they soak up most of the neutrons, so that a chain reaction cannot take place. What do you think would happen if a fast-moving chain reaction was not stopped? When the control rods are not all the way in, neutrons split the atoms and a chain reaction is started. Moving the control rods out starts and then speeds up the chain reaction. Moving them in slows it down and then stops it. What controls the chain reaction?

In reality, the two straws in the side of your shoebox reactor model would circulate water in and out of the nuclear reactor. The water has three purposes: (1) to help slow down some of the fast moving neutrons which might cause a chain reaction to happen too fast; (2) to deflect the neutrons that bounce out of the rods, and cause them to return to the rods; (3) to cool down the operation in the nuclear reactor. The temperature produced in the reactor is more than 500 degrees Fahrenheit, thus, the heat generated in the reactor is carried away by the water and is used to make electricity.

While a nuclear reactor is making energy, it is also generating wastes; it is making **radioactive** atoms. These atoms are very dangerous since the rays that come from them cannot be seen or felt, but if they strike a living body they can cause sickness and even death. To protect the people who work around these radioactive materials, special clothing must be worn. To protect us, the radioactive wastes are collected and stored in shielded containers where they will stay until they stop being radioactive. Some wastes are stored in tanks, others are buried in the ground.

More Questions About Fission and Fusion

Describe the differences between a fusion reactor and a fission reactor. How are they alike? How are they different? Which type do you think should be developed as our future energy source?

What does "radioactive decay" mean?

Before uranium can be used in nuclear reactors to produce energy, it must be mined, milled, refined, enriched, converted, and fabricated. Find out additional information about each of the six steps in the process from raw materials to nuclear fuel.

What are the advantages and disadvantages of "going nuclear"? What's your recommendation?

Nuclear waste disposal is a serious problem and a threat to our survival. Collect evidence to defend or refute our present nuclear waste disposal policies.

SOURCES OF ENERGY BINGO

Procedure

Each player cuts out sixteen squares (see Figure 9 for labels and numbering of the squares) and shuffles them. The players then select squares at random, one at a time, and glue them onto a piece of paper, forming their own four-column square bingo card. This will then serve as their personal *Sources of Energy* bingo card. It will be different from those of the other players. Choose a caller to cut up the clues (Figure 10), and put them in a container. The caller is the only person who doesn't have a bingo card.

When the caller draws a clue, he or she reads the one sentence description of an item related to a source of energy (never the corresponding number), and the players then place a marker over the correct term on their bingo cards. The first player to get four boxes in a row wins.

Check the answers aloud to make sure that the winning player has identified the items correctly. If the player is incorrect, he or she is eliminated from the round and the game continues.

FIGURE 9   SOURCES OF ENERGY BINGO CARD

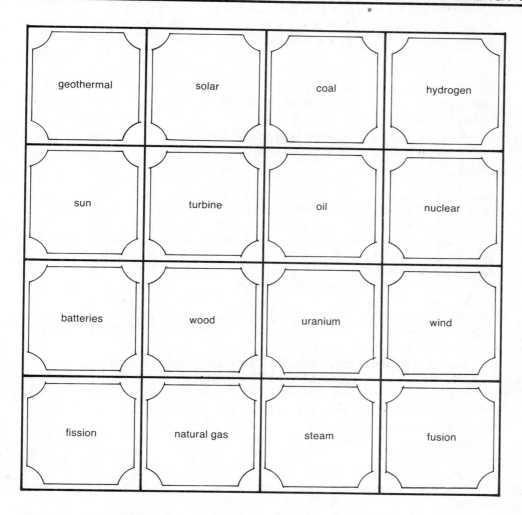

FIGURE 10   SOURCES OF ENERGY BINGO CLUES

Clues	Card Square Number
Heat from inside the earth	(1)
Nature's biggest energy source	(2)
Used for storing energy	(3)
Process for splitting the atom to produce energy	(4)

Process that will convert sunlight into energy	(5)
Turned by steam to produce electricity	(6)
Was once a major source of fuel	(7)
Energy source often found associated with oil	(8)
The primary energy reserve in the United States	(9)
The only fossil fuel in liquid form	(10)
The element which serves as the source for nuclear reactor fuel	(11)
Type of energy released when water is boiled	(12)
Could possibly serve as a future source of energy except for its high flammability	(13)
Excellent source of energy except for its waste disposal problem	(14)
Ancient source of energy that is presently being revived	(15)
The "clean" method for generating nuclear energy	(16)

More Energy Bingo Games

Construct other Energy Bingo Games utilizing topics such as: conservation, alternative sources, energy problems, fuel types, and so forth.

WHY CAN'T WE HAVE "FREE ENERGY"? PERPETUAL MOVING MACHINES?

The following five perpetual motion machines are designed to remain in motion after they have been placed in motion. However, all perpetual motion machines on earth are doomed to failure. Do you know why? All machines will slow down very quickly unless they have a constant supply of energy to overcome the effects of gravity.

Observe each machine and try to figure out why it will not "run forever" as a perpetual motion machine. The answers are given at the end of this activity, but don't spoil the fun by peeking!

Pivoting Balls

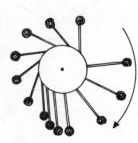

There have been many variations of this type of perpetual motion machine. As the wheel revolves, the weights are thrown out to the right, where they have greater leverage and therefore keep the wheel spinning. Each successive weight adds its thrust. Why won't it work?

A Cycle Powered by Magnetism

This is one of the simpler but more intriguing devices. A large round lodestone or magnet is supposed to pull an iron ball up the incline. Arriving at the hole, the ball should drop through, run down the trough, and out a trap door, ready to be drawn up again. Why does it fail?

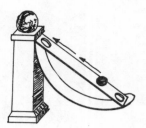

The Pumping Water Wheel

A pumping water wheel is operated by water dropping through a hole in the top reservoir. The water falls into the empty section of the wheel immediately beneath the hole, fills it until it gets so heavy it turns the wheel so that another section can be filled. The wheel drives a pump which lifts water from the bottom reservoir. The water is lifted up to the top reservoir, finds its way through the opening and turns the water wheel. What happens to prevent it from working perpetually?

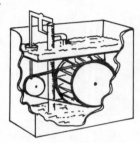

The Chain That Runs Forever

This is an endless chain mounted on pulleys. Free-rolling idler wheels on one side make that side of the chain longer than the other side. Since the chain is off-balance, the added weight on the right should pull the chain around and around. Why won't it work?

The Cage and Balls

Here is a leverage problem. The cage is divided into pockets, each of which contains a heavy ball (cross section of cage is shown). Because the balls at the left are farther from the hub than those at the right, the wheel should turn in a counter-clockwise direction. As it turns, more balls roll toward the rim to keep it spinning. Why does it fail as a perpetual motion machine?

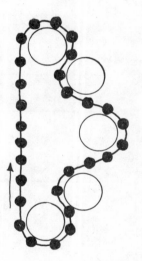

Why They Won't Work

The pivoting balls apparatus soon grinds to a halt because there are more weights on the left side, balancing the increased leverage of the weights on the right. This makes the wheel quickly reach a state of equilibrium. Recommend changes that would make it work longer.

The problem with the cycle powered by magnetism machine is that a magnet strong enough to pull the ball up the plane is too strong to let it drop through a hole again. How could this be rectified?

The pumping water wheel works when the water drops through a hole from the top reservoir and turns the water wheel which empowers the pump to push the water back up to the top reservoir. Unfortunately, friction won't permit the apparatus to lift up as much water as falls through the opening, so all the water eventually ends up in the bottom reservoir. Think of ways to improve this machine.

The chain that runs forever does not work because part of the weight of the right side of the chain is supported by the idler wheels at the points where the chain curves around them. This offsets the extra weight of the right side of the chain.

The cage and balls is an optical illusion. There isn't greater leverage at the left — the total force exerted downward on each side is exactly the same and the wheel will quickly reach a state of equilibrium.

Where could you travel with your perpetual motion machine so that it would work forever without any gravitational effect to disrupt it?

Design a perpetual motion machine. After you design one, try to build it!

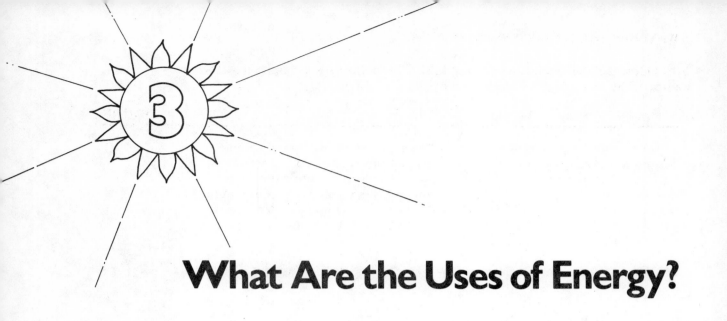

What Are the Uses of Energy?

We have applied energy to a wide range of uses—cooking, heating and cooling, lighting, transportation, hot water, refrigeration, communication, and clothes drying are a few.

Think of non-essential uses of energy that are for comfort and convenience. List the energy uses in your home that are essential and those that are non-essential in a table similar to the one we have started in Figure 1.

FIGURE 1

Essential	Non-essential
Heating	Can opener
Lighting	Clothes dryer

When do essential items become non-essential? When do non-essential items become essential? How do people from different parts (east, west, south and north) of the United States view essential and non-essential items? Extend this comparison to people who live in Canada, Mexico, England, and the United States.

Look through magazines and cut out pictures related to energy usage—cars, stores, refrigerators, airplanes, fans, etc. Prepare two large wall charts labeled "Things We Need" and "Things We Don't Need." Have the children decide which picture should be posted in which category. Discuss the reasons for their choices.

DOES IT MAKE SENSE?

The United States, with 6% of the world's population, uses 30% of the world's available energy.

DID YOU KNOW?

The United States uses more energy than West Germany, Japan, Great Britain, and the Soviet Union combined.

Observe the circle graph* in Figure 2 for the uses of energy in the mid-1970s.

FIGURE 2

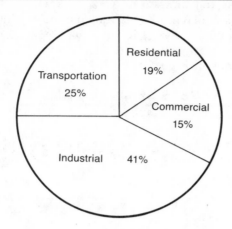

Residential
19%

Transportation
25%

Commercial
15%

Industrial   41%

*Source: *Encyclopedia of Energy*, Daniel W. Lapedes, Editor in Chief, McGraw-Hill Book Company, New York, 1976, p. 3.

Which segment of our society uses the most energy? Which segment uses the least energy? Infer reasons for this distribution of energy consumption.

Let us examine each of the four areas shown in our circle graph. First we will investigate the uses of energy in transportation. Energy use for transportation is shown in Figure 3.

FIGURE 3   ENERGY USE FOR TRANSPORTATION OF PASSENGERS AND FREIGHT*

Automobiles ————————————————————— 55%

Trucks ————————————————————— 20%

Airplanes ————————————————————— 10%

Barges and Ships ————————————————————— 5%

Fuel Pipelines ————————————————————— 5%

Railroads ————————————————————— 4%

Buses ————————————————————— 1%

*Source: *Energy and Society: Investigations in Decision Making*, Biological Sciences Curriculum Study, Hubbard Scientific Company, P.O. Box 104, Northbrook, Ill. 60062, 1977, p. 63.

Which mode of transportation is used the most? Why do you think this is? Which mode of transportation is used the least? How can you account for this?

Examine the modes of transportation in your community. Which method of transportation is used the most? Which method is used the least? Does your community have mass transportation available such as the subway, buses, monorail, and so forth? Which method of transportation, airplane, truck, ship or train, do you think would be most efficient in moving materials from one point to another? Rank them from most to least efficient. How do your rankings compare to the rankings in Figure 4?

FIGURE 4   HOW FAR CAN THIRTY TV SETS BE MOVED ON A LITER OF GAS?*

Airplane _____ 1 km

Truck _____17 km

Ship _____71 km

Train _____72 km

*Source: *Energy and Society: Investigations in Decision Making,* Biological Sciences Curriculum Study, Hubbard Scientific Company, P.O. Box 104, Northbrook, Ill. 60062, 1977, p. 64.

What is the primary method of transporting material in your community? Suggest ways to improve the transportation methods used in your community.

Next, consider ways of moving people (passengers) in the city and between cities. Rank the inter-city transportation methods from most to least efficient. Rank the city to city transportation methods in the same manner. Did you choose the train, bus, airplane, or automobile as the most efficient in each case? Compare your results with Figure 5.

FIGURE 5   HOW FAR CAN ONE PASSENGER BE MOVED ON A LITER OF GAS?*

In the City		Between Cities	
Auto _____ 7 km		Auto _____16 km	
Bus _____14 km		Bus _____33 km	
Train _____18 km		Train _____18 km	

*Source: *Energy and Society: Investigations in Decision Making,* Biological Sciences Curriculum Study, Hubbard Scientific Company, P.O. Box 104, Northbrook, Ill. 60062, 1977, p. 64.

To graphically illustrate the information presented in Figures 4 and 5, mark off an area of a floor or playground using one of the following scales: 1

centimeter equals 1 kilometer; 1 meter equals 1 km (kilometer). Then, mark off the distances shown in each table.

Automobiles are, without a doubt, a wasteful form of transportation. Statistics show that more than half of all auto trips are for less than 5 miles, and three-quarters are for less than 10 miles. Usually these trips are trips to work or to the store, and the driver is generally alone in the car. The increased size of American cars increased their inefficiency. A car weighing 5,000 pounds uses over twice as much fuel as a car weighing 2,000 pounds. Extras such as air conditioning, a larger engine than needed, and power equipment further decrease the mileage per gallon.

Try the following automobile related activities.

COME AWAY WITH ME IN MY MERRY...

Procedure

Take a paper and pen and station yourself on a major street intersection for a set amount of time (15 minutes at a busy time of day). Keep a record, based on the following criteria, of what you observe.

> type of cars—subcompact, compact, medium-sized, large, and luxury
> number of people per car—1, 2, 3 . . .
> number and type of cars passing per minute

Next, keep a record of your observations based on the following questions.

> How many one-person cars pass per minute?
> How many two-person cars pass per minute?
> How many cars containing more than two people pass in a minute?
> Is there a relationship between the type of car and the number of people inside?
> Which type of car is most prevalent?
> Which type of car is least prevalent?
> What explanation can you offer for this?

DID YOU KNOW?

An electric car can go up to 40 miles in one day before it needs to be recharged.

DOES IT MAKE SENSE?

Does it make sense to take a 5,000 pound car around the corner for a quart of milk at the grocery store?

Check an automobile's mileage. To do so, children will have to choose a car that belongs to someone who will cooperate with them in the experiment—parents, a teacher, or a friend.

What To Do

- Write down the year, model and name of the car you're using.
- Record the odometer reading when the gasoline tank has just been filled.

- Fill the tank again when the gas gauge reads one quarter full and record the odometer reading, the number of days elapsed between fill-ups, and the number of gallons (or liters) or gasoline used.
- Subtract the first odometer reading from the second to determine the number of miles travelled.
- Compute the number of miles per gallon (or liter) for the car by dividing the number of miles travelled by the amount of gasoline used.
- Compute the average number of miles travelled per day by dividing the number of miles travelled by the number of days between fill-ups.

DID YOU KNOW?   World's Petroleum Reserves (Proven Recoverable)

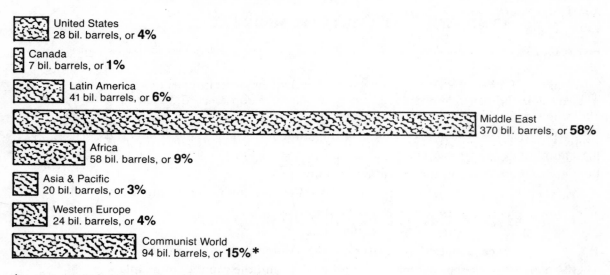

United States
28 bil. barrels, or **4%**

Canada
7 bil. barrels, or **1%**

Latin America
41 bil. barrels, or **6%**

Middle East
370 bil. barrels, or **58%**

Africa
58 bil. barrels, or **9%**

Asia & Pacific
20 bil. barrels, or **3%**

Western Europe
24 bil. barrels, or **4%**

Communist World
94 bil. barrels, or **15%** *

*Includes some potentially recoverable oil.

If each class member obtains the data needed above, the following comparisons can be made:

- Compare highway mileage with city mileage.
- Compare subcompact, compact, and large cars with respect to mileage.
- Compare different years of the same make of car. Are mileages improving?
- Compare different models of the same car—2-door, 4-door, hatchback, station wagon. Any differences?

More

You can extend this experiment by checking how mileage is affected by the number of passengers in the car; varying weather conditions; or a diesel versus a gasoline engine in the same make, year and model car.

The efficiency of a gasoline powered engine is only about 20% while the efficiency of a battery powered engine is about 90%. The energy contained in one gallon of gasoline is approximately equivalent to the energy contained in fifty batteries. Thus, if the battery is about four and a half times as efficient as a gasoline powered motor, you would need eleven batteries to provide the energy of one gallon of gasoline. Therefore, you would need approximately two batteries to provide the energy needed to travel 20 miles per day.

Obtain a battery-powered, toy car. How many batteries does it need to work? How long will the toy car run until it needs new batteries? What are the advantages of electric cars versus gasoline powered cars? What are the disadvantages?

DOES IT MAKE SENSE?

Eight 6-volt batteries are needed to operate a 1500-pound electric car. The batteries must be replaced every 15,000 miles.

DOES IT MAKE SENSE?

Almost 87% of the energy gasoline powered automobiles use is shot out of the exhaust pipe.

Find out if there is an electric car distributor in your area and see if your class can visit it or if you can arrange for someone with an electric car to visit your school (a list of electric vehicle manufacturers is provided in the bibliography). Observe, compare, and contrast the car with a gasoline powered car of similar size (a subcompact). Have each student decide what conditions would be beneficial to people using an electric car, and the conditions that wouldn't be conducive to using an electric car.

Other Forms of Transportation

Airplanes are also wasteful. They use twice as many Btu's of energy per passenger mile as cars, and six times as many as trains. The number of flights scheduled by airlines in an effort to bring in more business makes this waste even greater. Visit a nearby airport and determine the number of people per plane versus the number of available seats. Also find out what airlines and airline manufacturers are doing to conserve fuel. Figure out some ways for the airlines to conserve fuel, like elimination of duplicating flights and reducing airplane speeds.

Air and truck transport use up 29% of all transportation energy. It has been estimated that measures to consolidate trucking and to encourage railroad freight transport could save more than 4% of expected transport energy. Invite a local trucking industry representative to visit your classroom. What efforts are they making to reduce their energy consumption?

There appears to be a concerted world-wide effort to utilize mass transit and inter-city trains so that people will have an alternative to automotive and air transport. What are the advantages of relying on mass transit? What are the disadvantages? Compare the effect of mass transit on rural areas versus urban areas. Survey the people in your community. Do they utilize mass transit now? Will they in the future? Why or why not?

TRANSPORTATION TOGETHERNESS

On one side of the room, group five chairs in a long line to resemble seats on a bus. Put five chairs on the other side of the room; keep them all separate to represent five separate cars. Pick five people to act out riding on a bus together,

and choose another five to pretend they are all driving to the same place in their separate cars. Why is one form of transportation better and more efficient? What are the advantages and disadvantages of using car pools to go to and from school?

Using magazines, cut out pictures of every kind of car that you can find. Rank them from smallest to largest. What are the similarities and differences between the cars? Which ones would use the most gas? Which ones have the best mileage? Which ones look the nicest? Which ones cause the most pollution?

Show your class a picture of a modern car. Ask everyone to locate pictures of a form of transportation older than the car (such as an older car, or a horse and buggy, a horse, etc.). Then, ask everyone to draw their idea of the best transportation method of the future—you might want to choose a year like 2042. Discuss the advantages and disadvantages of each proposed future method of transportation.

Household and Commercial Uses of Energy

APPLIANCE RELIANCE I

To understand the reliance we place upon energy for household and commercial uses, fill in Figure 6 either as a group or individually. Place an "X" in the column or columns to indicate the type of energy source for each energy user listed. Some appliances or machines can operate on more than one type of energy. The answers will depend on the energy sources in your area.

DOES IT MAKE SENSE?

Does it make sense to use trucks for inter-city freight when railroads, which handle 36.2% of the nation's inter-city freight, use only 3.3% of the petroleum consumed by the transportation industry?

FIGURE 6   SOURCES FOR ENERGY USERS

Energy User	Electric Battery	Gas	Electric Current	Nuclear	Coal	Oil	Wood
Furnace							
Clock							
Television							
Refrigerator							
Fireplace							

APPLIANCE RELIANCE II

Suppose you wake up one morning and discover that you will be limited to 3500 watts of electricity from now on. Examine the list of appliances in Figure 7, and determine the wattage each requires. Then decide which of them you would choose to use and enter them in Figure 8. Remember, you are limited to 3500 watts total.

FIGURE 7 AVERAGE APPLIANCE WATTAGES

Appliance	Average Wattage
Window air conditioner	1500
Clock	2
Clothes dryer	5000
Coffee maker	900
Dishwasher	1200
Food blender	400
Food freezer (15 cu ft)	350
Food freezer (frostless, 15 cu ft)	450
Food wastes disposer	500
Frying pan	1200
Hair dryer	400
Radio	70
Electric range	12,000
Refrigerator-freezer (14 cu ft)	300
Refrigerator-freezer (frostless 14 cu ft)	600
Stereo	110
Television (B & W)	250
Television (color)	350
Toaster	1200
Washing machine (automatic)	500
Water heater (standard)	2500
Water heater (quick recovery)	4500

FIGURE 8 MY APPLIANCE CHOICES (LIMIT 3,500 WATTS)

Appliance	Average Wattage

DOES IT MAKE SENSE?

A frost-free refrigerator uses 39 to 50% more electricity so that you will not have to defrost it every six months.

DOES IT MAKE SENSE?

A recent home survey found thirty-two electric appliances in the kitchen, twenty in the dining room, and twenty-nine in the bedroom.

After completing Figure 8, compare your choices to others. Why did certain people make certain choices? What do these choices indicate about one's life style and values? Try this same activity with your friends and your parents. Are the results the same? Why or why not? Would they have selected the same appliances five years ago? Are there ways to eliminate the use of some of the appliances to save watts to apply to other appliances? Are there appliances that do things that you cannot do with your own hands?

Next, on a piece of paper titled, "How I Use Electricity," draw pictures of everything that you used this week that required electricity. Rank the appliances on the list from most important to least important. Decide which items on the list could be eliminated or their usage reduced. Which items are essential and could not be reduced or eliminated?

Find and laminate pictures of "energy usage" rooms (such as a kitchen, family room, laundry room, etc.). Pass these pictures around with grease pencils and ask everyone to circle the "energy users" they find. Next, ask them to "X" out energy users that could be eliminated. Discuss the advantages and disadvantages of everyone's choices.

Group the appliances in the following list into categories of appliances with heating elements, and those without heating elements.

Broiler	Floor polisher
Central heating unit	Food blender
Clock	Frying pan
Clothes dryer	Iron
Deep-fat fryer	Sewing machine
Dishwasher	Shaver
Fan	Vacuum cleaner

Write the wattage of each appliance on the right side of the column. Which appliances use more watts—those with heating elements or those without? Which group should you reduce first? Why?

Find out which electrical appliance in your home has the highest wattage rating and which has the lowest. Did you think that it would be the appliance you first thought of? Why or why not?

DID YOU KNOW?

In 1931 Americans spent 1 billion dollars on appliances. Just 50 years later in 1981 they will spend 20 times that amount.

DID YOU KNOW?

A solid state color television set uses 20% less energy than a regular color television set.

Prepare a list of the rooms in your house then write the number of outlets each one contains. Next, count the number of appliances used in each room. Which room has the most appliances? Which room has the least? Which appliances could "never" be unplugged? Which appliances should never remain plugged in? Try going without an "essential" appliance for an hour, a day, a week. How long could you go before you discovered you indeed needed it?

Which appliances do you have now that weren't available thirty years ago, ten years ago, or two years ago?

Design an appliance that has not been invented yet. Then decide if your new appliance is really needed.

Do a pictorial essay of what life was like in 1888, before electricity was used. Then do a drawing or picture collection of electricity today. Finally, draw a picture of what you think our electrical future will look like in the year 2042.

Energy usage in the average home is:

57% space heating	15% water heating
4% air conditioning	2% clothes drying
6% cooking	2% food freezing
6% refrigeration	8% other

Obtain an average month's electric bill for your home and calculate how much each area of the home costs. To do this, you'll have to figure out the amounts of electricity used in each room, assign each room a percentage figure (of the overall use of electricity in the house), and divide the bill up accordingly. Compare your results with other people's results. Indicate where the greatest savings might occur.

To estimate the cost of running a room air conditioner in your room, collect the following data: the cooling-season electric rate in cents per kilowatt-hour and the average number of hours of cooling required in a season (both figures are available from your electric power company); your cooling load (an average size room is 6,500 Btu/hr), and the labeled energy efficiency ratio (EER) of the unit you are considering. A sample calculation is shown in Figure 9.

FIGURE 9   COST OF RUNNING A ROOM AIR-CONDITIONER

Calculation	Our Example	Your Room
Electric rate times the hours of cooling	10¢/kWh times a 700 hour cooling season = 7,000	___ × ___ = ___
Item 1 times your cooling load	7,000 × 6,500 Btu/hr = 45,500,000	___ × ___ = ___
Drop the last 5 digits from item 2	455	___
Annual operating cost (item 3 ÷ EER)	455 ÷ 8.7 = $52.30	___ ÷ ___ = ___

The average electric fan costs $11.50 to run for seven hundred hours over a summer's cooling season and can be purchased for $50.00 compared to $300.00 for a window air-conditioner. Prepare a comparison chart for fans versus air-conditioners which cites the advantages and disadvantages of each. How can you decide which to purchase and use for cooling purposes? Interview people who use fans and those who use air-conditioners. What are the reasons cited by these two groups for choosing one over the other? Is it cost, noise-level, comfort, or what?

Industrial Uses of Energy

Industry is another energy use sector that affects our daily lives, but its influences are often overlooked because we tend to take them for granted. Nearly everything in your home and school has been manufactured or processed. The paper in this book and the ink for its printing are products of industry. Observe your classroom. What would be left if all the industrial products used in it were suddenly not available? The major energy uses for industry are for steam and direct heat. Figure 10 shows the major types of industry and the relative proportions of their energy use.

DID YOU KNOW?

One hundred years ago it took about the same amount of energy to heat a house as it does today. It took about half as much energy to feed the family horse as it now takes to power the family car. It took about the same energy to cook a meal.

FIGURE 10   INDUSTRY ENERGY USE*

Industry	Percent of Energy Use
Chemicals	17%
Metals	17%
Products from fossil fuels	9%
Blast furnaces and steel mills	8%
Paper	7%
Food	7%
Stone, clay and glass	6%
Petroleum refining	6%
Others	21%

*Source: *Energy and Society: Investigations in Decision Making*, Biological Sciences Curriculum Study, Hubbard Scientific Company, P.O. Box 104, Northbrook, Ill. 60062, 1977, p. 65.

Note that 9% of industrial use of energy is devoted to products made from fossil fuels. These are industries where coal, oil, and natural gas are used, not as energy sources, but as raw materials. Synthetic fibers, plastics, and some drugs are made from fossil fuels. These products are extremely important to us and, if we ever run out of fossil fuels, they will no longer be available to us. Thus, one valid reason for fossil fuel conservation is to preserve much needed raw materials.

WHO USES WHAT, WHEN, WHERE, AND WHY?

Identify which of the industries in Figure 10 are located in your area. Which industries use fossil fuels as raw materials? Which industries use the most energy? Have a representative from selected industries tell your class what they are doing to utilize energy more economically. Try to arrange a visit, if possible, to that industry so they can directly observe proper energy management. After the visit, discuss how energy played a part in the operation of the plant. Discuss a product that the group could manufacture and try it. Maybe you would like to mass produce cookies or bird houses. Try making the product with no outside energy source, with essential energy only, and with lots of energy. Discuss the advantages and disadvantages of each method.

Present pictures or drawings of finished products (for example: car, house, bike) to each other. Discuss how that finished product came to be and the steps involved in making the product.

Many people do not realize that the biggest user of natural gas is industry (45% of the total demand). In addition to its use as a plant fuel, industry uses natural gas in the processing of metals, crude oil, and other raw materials; in the processing of food products; and for many other purposes.

DID YOU KNOW?

Wise use of energy by every large United States corporation could save enough energy to equal the yearly output of twenty-four average-size oil refineries.

Uses of energy in industry can be made more efficient through three phases as shown in Figure 11.

FIGURE 11 PHASES OF INDUSTRIAL ENERGY USAGE IMPROVEMENT*

Phase	Potential Savings
1. Housekeeping	3-7%
2. Re-equipping existing processes	5-10%
3. Process change	10-90%

*Source: 1977 Purdue University Energy Conference Proceedings, West Lafayette, IN., p. 93.

Housekeeping (Phase 1) refers to the typical energy savings from turning off unneeded lights, unused motors, plugging air leaks, etc. Re-equipping existing processes (Phase 2) involves minor process changes such as installing some basic improvements that are not extremely costly. Major process changes (Phase 3) cannot simply be justified by a lower fuel bill, but involve a basic economic reason for replacing a major existing production facility with a new facility. This change will not occur over a short period of time, but will take from twenty-five to fifty years, or more.

Identify the three phases of industrial energy usage improvement in your area. Are all three phases in evidence? Why or why not?

Pretend your school is an industrial building. Prepare a list of suggestions to improve energy usage by completing the phase chart in Figure 12.

FIGURE 12 PROPOSED PLAN TO IMPROVE ENERGY USAGE IN OUR SCHOOL

Phase	Recommendations	Potential Savings
1. Housekeeping 2. Re-equipping existing processes 3. Process change		

Energy Use in the Food System

Energy inputs to farming have increased enormously during the past fifty years, and the decrease in farm labor is offset in part by the growth of support industries for the farmer. A variety of other changes have transpired in the United States food system, many of which are now deeply embedded in the fabric of our daily lives. In the past sixty years, canned, frozen, and other processed foods have become the principal items of our diet. Presently, the food-processing industry is the fourth largest energy consumer in the United States. Transportation within the food system has steadily increased to match demand. Appliances of varying complexities continue to be invented and bought for use in homes, institutions and stores. Hardly any food is eaten as it comes from the fields. Even farmers purchase most of their food from markets in town.

A typical breakfast includes: orange juice from Florida, by way of a factory; bacon from a midwestern meat packer; cereal from Nebraska and a cereal factory; eggs and milk from a nearby farm; and coffee from Colombia. All of these things are available at the local supermarket (only several miles away and quickly reached by a 300-horsepower automobile), are stored in a refrigerator-freezer, and cooked in an instant-on stove.

WHAT USES UP ALL THAT ENERGY?

In exploring energy use in the food system, we will begin with an omission. We will neglect the crucial input of energy provided by the sun to the plants upon which our entire food supply depends. Photosynthesis has an efficiency of about 1%. Thus, the maximum solar radiation captured by plants is about 5 by 10^3 kilocalories per square meter per year. Almost 50% of the total use of energy in agriculture is for growing food. Food processing is the next major energy user. (See Figure 13.)

FIGURE 13   ENERGY USE IN THE FOOD SYSTEM*

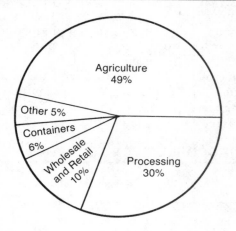

*Source: 1977 Purdue Energy Conference Proceedings, p. 130.

Figure 14 indicates the uses of energy on the average farm. Direct use of fuel uses almost half the energy in farming. In the United States only 5% of the crop land is irrigated; note that irrigation accounts for 5% of the energy usage even though the irrigated land usage is quite low. If the United States did not have such superior farm land and had to irrigate a substantial portion of land for new high-yield varieties of plants, irrigation would be the largest single use of energy on the farm.

Very little food makes its way directly from the field and farm to the table. The vast complex of processing, packaging, and transporting has been grouped

FIGURE 14 ENERGY USAGE ON THE FARM*

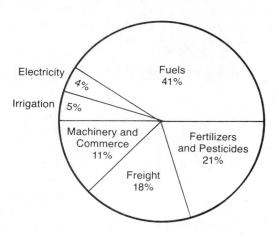

*Source: 1977 Purdue Energy Conference Proceedings, p. 131.

together in a second major subdivision of energy use in the food system (see Figure 13). After the processing of food, there is further energy expenditure. Transportation enters the picture again and there are also the distributors, wholesalers, and retailers whose freezers, refrigerators, and very establishments are an integral part of the food system. There are also the restaurants, schools, universities, prisons, and a host of other institutions engaged in the procurement, preparation, storage, and supply of food. Energy is also used in trips to the store or restaurant. Garbage disposal is a persistent and growing feature of our food system. Twelve percent of our nation's trucks are engaged in the energy activity of waste disposal.

HOW CAN WE USE LESS ENERGY?

Pimentel and his associates* have made five suggestions to reduce the energy usage required for agriculture and the food system. They are:

1. *Make more use of natural manures.* The United States has a pollution problem from runoff from animal feed lots.
2. *Weed and pest control could be accomplished at a much smaller cost in energy.* Changes in cultivation habits and biological pest control (such as the use of sterile males and introduced predators) requires only a fraction of the energy of pesticide manufacture and application.
3. *Plant breeders will need to breed greater hardiness, disease and pest resistance, reduced moisture content (to end the use of natural gas to dry crops), reduced water requirements, and increased protein content in plants.*
4. *Chemical farming should be abandoned.*
5. *The direct use of solar energy farms, a return to wind power, and the production of methane from manure should be explored.*

*Source: D. Pimentel, L. E. Hurd, A.C. Bellotti, M. J. Forster, I. N. Oka, O. D. Scholes, R. J. Whiteman, *Science* 182, 443 (1973).

Divide into five groups and have each group select one of the five energy usage reduction proposals presented. Each group should further investigate their proposal and develop a set of positive and negative effects that would occur if the United States adopted their proposal. To illustrate, we have proposed one, as shown in Figure 15.

FIGURE 15   CHEMICAL FARMING SHOULD BE ABANDONED

Positive Effects	Negative Effects
Less chemical pollution.	Reduced crop yields per acre.
Most land in the soil bank would be placed back into production.	Output would fall 5%.
Farm income would rise 25%.	New crop diseases might develop.
All governmental subsidy programs would be ended.	Insect pests might take the upper hand in crop management.

FOOD GENEALOGY

Select a food product and trace it back to its source, the way we did with the bacon and orange juice. Try to arrange a field trip to a local farm, food processor, and supermarket to illustrate how energy is used to bring food from the farm to our home.

Next, prepare a diorama illustrating the path that food takes from the farm to the home and the energy users in each step.

See if the representative responsible for your school's lunch program will visit your classroom to talk about how the food travels from the farm to the school. Identify the energy users. Find out what steps are being taken by the school's food program to reduce energy usage. Make a "behind-the-scene" visit to your school lunch preparation area and identify areas of large and small energy usage such as the refrigerators and ovens compared to clocks and lights. What can be done to reduce energy usage?

Interview children in the school lunch line to find out their suggestions for improving energy usage at lunch time.

Visit a dairy barn and observe the electric milking machines. Have the farmer demonstrate hand milking and then explain the advantages and disadvantages of each system.

Energy from Foods

It is interesting to note that high-protein foods such as milk, eggs, and especially meat have a far poorer energy return than plant foods. Soybeans possess the best amino-acid balance and protein content of any widely grown crop.

Find out more about soybeans as a food source. Locate products in the supermarket made from soybeans and try them. Which product do you prefer and why? Bring in a soybean plant and compare it to other plants. Why is it so nutritious and suited to wise energy usage?

Energy Usage and Electric Power

A fuel is a substance burned to create energy. The "fossil fuels" (coal, oil, and natural gas) power most of our offices, homes, industries, and transportation. And, since the discovery of electricity, the fossil fuels have been used to produce electric power as well. Many people think of electricity as a fuel. It is not. Fuel materials such as coal, oil, gas, and uranium are converted to electricity in power plants. These fuels are burned to make heat and this heat warms water in huge boilers. The steam from the boiling water sprays onto the blades of a huge fan, or turbine, which is attached to a magnet. The spinning magnet creates electricity in coiled wires. The magnet and wires are called a generator. Electricity is the biggest business in this country and its use is expected to grow about seven times in the next thirty years. Seven times more electricity would allow a growth of about 600% in household use of electricity, 1,100% in commercial use, and 300% in industrial use. The steps from fuel to electricity are shown in Figure 16.

WHAT POWERS YOUR SCHOOL?

Find out what type of fuel is burned in the electrical power plant that your school obtains electricity from. Prepare a map showing the location of the power plant and how the electricity is transported to your school. Arrange a field trip to visit your local electric power plant. Have everyone relate what they observe to Figure 16. Take a series of photographs or cut out pictures of electric power generation and arrange them in order from the beginning source of fuel to electrical power transmission.

There are three significant types of generating plants that convert some form of energy to bulk electrical energy: hydro-electric, fossil-fuel electric, and nuclear electric plants.

The hydro-electric plant utilizes the potential energy released by the weight of water falling through a vertical distance. The plant basically consists of a dam to store the water, a method to deliver falling water to the turbine, and a generator to convert the mechanical energy to electrical energy. Pumped storage hydro-electric plants are being increasingly used. During non-peak hours (10 P.M. to 6 A.M.) the surplus generating capacity is used to return the water to the area behind the dam for reuse during peak hours.

DID YOU KNOW?

The waste of one cow equals the waste from sixteen people; the waste from one hog equals the waste from two people; and the waste from seven chickens equals the waste from one person.

Fossil-fuel electric plants utilize coal, oil or natural gas. A typical large plant consists of fuel processing and handling facilities, a combustion furnace and boiler to produce and superheat the steam, a steam turbine, an alternator (generator), and the accessory equipment required for plant protection and for control of voltage, frequency, and power flow. Environmental standards require careful control of emissions through the smokestacks with respect to sulfur oxides and particulates. Cooling towers or ponds are often required for waste heat dissipation. Gas turbine plants do not require condensor cooling

FIGURE 16   FROM FUEL TO ELECTRICITY

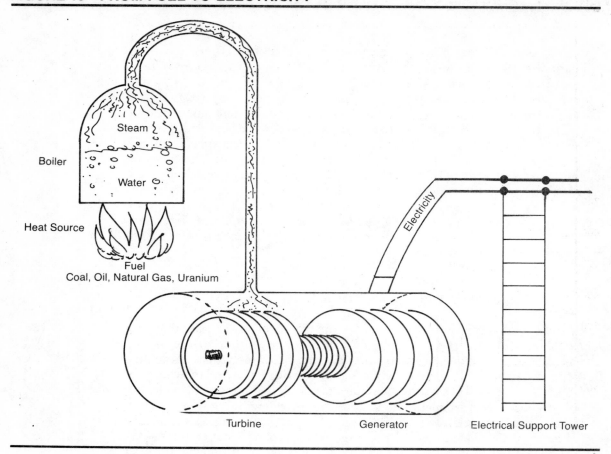

water, but do have a relatively high unit-fuel cost; thus they are used mainly for peaking service.

Nuclear electric plants utilize one or more nuclear fuels in a suitable type of nuclear reactor which takes the place of the combustion furnace in the typical steam electric plant. The rest of the parts of the nuclear electric plant are largely the same as the fossil-fuel electric plant.

As indicated in Figure 17, the generating capacity of nuclear power plants is increasing (estimated to be 25% of total generating capacity by 1985), hydro-generation reached a saturation point in 1975 and fossil-fuel generation has leveled off.

METER READING

What are the advantages and disadvantages of relying on one of the three types of electrical generating plants? What factors may prevent nuclear power plants from assuming 25% of the total generating capacity by 1985?

The watt is the common unit of power. Generally wattage is printed on light bulbs and appliances. It indicates how fast the light bulb or appliance uses electrical energy. If you know how fast a car traveled and you multiply that speed by how long it traveled, you can tell how far it went. In the same way, if you know how fast energy was used and you multiply that speed by how long it was used, you can find how much energy was used.

For example, a 150-watt bulb burning for 60 seconds uses 9,000 watt-seconds of energy. To measure energy in your home, business, and schools, we

FIGURE 17   ELECTRIC GENERATING CAPACITY IN THE UNITED STATES BY TYPE*

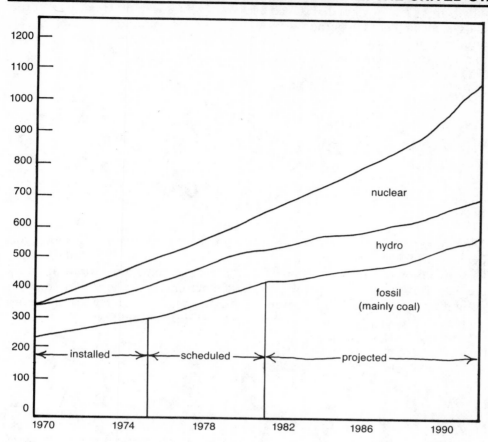

*Source: *Encyclopedia of Energy*, p. 201.

need a larger unit. In one hour, a 1,000-watt appliance uses a kilowatt (1,000 watts) hour of energy which is abbreviated kWh. The kilowatt hour is used in the electric meter in your home to measure the electricity used. Electric meters do not all look the same. Some can be read by simply reading the numerals showing like this:

00341

Others have four or five dials that need to be read.

Thousands

Hundreds

Tens

Ones

First observe the dials. Do you notice anything different about each dial? That's right, every other dial is read backwards. This is due to the way that the dials are connected together by gears. When the pointer is between two numbers on the dials, the reading taken is the lower of the two numbers. Thus the meter reading is: 5,560.

Now try this one:

Did you get 3,139? If not, make up your own meter reading problems until you become proficient. For five dials, the meter is read the same way, only you will have five numbers instead of four. Gas meters may have more dials, but they are read the same way. Prepare a large model meter to practice with before you go home to read your house meters. To find out how much electrical energy each person uses per month, let's practice reading meters first.

June 1

July 1

What is the meter reading for June 1?

Did you record 8,459?

What is the meter reading for July 1?

Did you record 8,758?

To find out how many kilowatt hours were used during June, subtract the June reading from the July reading.

$$
\begin{array}{r}
8{,}758 \text{ kWh reading for July 1} \\
\underline{8{,}459 \text{ kWh reading for June 1}} \\
299 \text{ kWh used during June}
\end{array}
$$

Now, you are ready to find out how much electricity you used in one month. Ask your parents where the meters for your house are located and take a reading.

Fuel oil, bottled gas, and gasoline are sold by the cubic foot. It is easy to figure the cost of energy from these sources, since it is usually so much per gallon or cubic foot. Electricity, however, is priced differently. Most electric companies charge a "sliding rate" for electrical energy. The cost per unit goes down as you use more electricity. Figure 18 shows the sliding rate for residential electricity charged in a city in Indiana. Find out your electric cost and prepare a chart similar to ours for use in calculating the amount of energy everyone in the class used.

DID YOU KNOW?

In 1975 only 8.5% of the total electrical generating capacity was nuclear. By 1985 it is predicted to be more than 25%.

DID YOU KNOW?

Only 14% of the young adults in America know that coal is the primary energy source used to produce the largest portion of the nation's electrical energy.

FIGURE 18   SLIDING RESIDENTIAL RATE ELECTRICITY CHARGED

First 12 kWh per month	— 28.7 ¢ per kWh
Next 88 kWh per month	— 7.2 ¢ per kWh
Next 100 kWh per month	— 5.3 ¢ per kWh
Next 400 kWh per month	— 3.6 ¢ per kWh
Next 400 kWh per month	— 3.4 ¢ per kWh
Over 1,000 kWh per month	— 2.6 ¢ per kWh

Suppose your meter shows that you used 1,524 kWh during the past month. This is how your bill would be figured:

First	12 kWh	12 × 0.287	=	$ 3.44
Next	88 kWh	88 × 0.072	=	$ 6.34
Next	100 kWh	100 × 0.053	=	$ 5.30
Next	400 kWh	400 × 0.036	=	$14.40
Next	400 kWh	400 × 0.034	=	$13.60
Remaining	524 kWh	524 × 0.026	=	$13.63
Total bill for 1,524 kWh used			=	$56.71

In addition to the charge of $56.71 for electricity, some companies add a fuel adjustment charge (which is a charge for increased fuel costs which is passed directly to the consumer) and state sales tax.

Calculate your electric bill for the past month using your rates. Does your electric company add a fuel adjustment charge? Does your state charge sales tax on electricity used? How is the fuel adjustment charge figured by your electric company? Some electric companies figure the fuel adjustment charge as a percentage of the electrical energy used, while others add a certain amount for each kilowatt hour used. For example:

1,524 kWh cost $56.71 in our example. If the fuel adjustment charge is 25%, this is how the fuel adjustment charge would be figured:

$$\$56.71 \times 25\% = \$14.18$$
$$\$56.71 + \$14.18 = \$70.89$$

Thus the cost of electrical energy ($56.71) plus a percentage of this amount ($14.18) gives the total bill ($70.89).

Suppose the fuel adjustment charge is one cent per kilowatt hour ($0.01/kWh). Then:

$$1,524 \text{ kWh} \times \$0.01/\text{kWh} = \$15.24$$
$$\text{Thus: } \$56.71 + \$15.24 = \$71.95$$

The cost of electrical energy ($56.71) plus the additional charge ($15.24) figured on one cent per kWh (1,524) gives the total bill ($71.95).

As energy needs increase and fuel becomes more scarce, it will cost more and more. Remember, you pay for all energy you use: the more you use, the more you pay; the less you use, the less you pay!

DID YOU KNOW?

The Department of Energy and the National Aeronautics and Space Administration (NASA) are using a 2-megawatt (mw) wind turbine to provide electricity for the local power-grid near Boone, North Carolina. The 350-ton giant turbine is on a 140-foot tower and is designed to operate at wind speeds between 11 and 35 miles per hour. Its twin 100-foot blades spin at a constant 35 revolutions per minute (rpm) and drive an alternating current (A.C.) generator at 1,800 revolutions per minute (rpm). This six million dollar unit is said to be able to provide enough energy to supply nearly 300 to 500 average size homes at a wind speed of 25 mph.

How Can We Conserve Energy?

Who Needs Energy?

Energy is the ability to do work. The work of energy produces power, heat, and light. The human body needs this energy to function. Running, working, even sleeping all require energy. Humans get their energy from food. For most of life, animals and plants alike, energy means food. The sun is credited as the ultimate source of our energy.

Machines also need energy to move and to do useful work. Most machines get power by using energy obtained from fossil fuels, such as coal, natural gas, and oil. These fossil fuels are also indebted to the sun as the source of their formation. Regardless of how the trail of energy turns and twists and on occasion doubles back, ultimately it all began with our stellar performer, the sun.

Energy is neither created nor destroyed. Energy exists. We capitalize on it. We need energy to accomplish the work of staying alive. Only its form is changed as various processes go on. For example, chemical energy is converted to thermal (heat) energy when fuel is burned. The amount of energy in the system is unchanged. Thus, the energy used is not really consumed, it is changed into another form. Although we don't consume the energy in a system, we consume the work accomplished by energy during this interchange process. In this process, energy is converted to another form which usually eludes our further use, but nevertheless, it is there in one form or another.

DOES IT MAKE SENSE?

The best overall measure of the capacity for doing any task is the work—the transfer of the highest quality energy from one system to another. For every two units of energy that go into America's entire energy grid, only one unit of useful work comes out. (That is a 50% efficiency rate!)

Progress requires energy. While all living things need air, water, and food to stay alive, humans seem to have requirements for more than these simple basics. Progress brings with it a requirement for more—more clothes, more cars, more medicines, more food, and so forth. The more successful we become, the more things we seem to need. We desire bigger and better things.

We are truly creatures of comfort. We like it warm indoors when it is cold outdoors. We like it cool indoors when it is hot outdoors. When the humidity is too high, we want it lowered. When it is too low, we want it higher. When it is dark, we want light. When we eat, we want our foods broasted, toasted, baked, fried, stewed, brewed, and so forth. We like our chocolate, tea, or coffee hot, and our milk cold. We like to stay in the shower until the hot water runs cold. We want to be surrounded by sound, either stereophonic or television. We want to wash and dry our clothes in minutes. And, we want to travel and to make phone calls when and where we like. We use a large amount of energy to make our lives comfortable. How much energy can one save and still not grossly deteriorate one's life style?

Describe how human consumption is related to human comfort. What would be an average daily minimum energy comfort requirement expressed in wattages for each member of your family?

WATTS DOWN IS WHAT'S UP*

We all have a part to play in the current, and what may well be persistent, energy crisis. Some of us have been living "energy" fat. We use too much coal. We use too much wood. We use too much gas. We use too much oil. And we use too much electricity. Some things we can control; some things we cannot. We can all exercise control in the amount of household electricity we use. Electricity is usually generated by burning oil or gas. Thus, if we waste electricity, we are wasting gas and oil.

In the 1930s a typical American household had twenty to thirty appliances. In the 1940s this figure increased to thirty or forty appliances per household. By the 1950s, this statistic had increased from forty-five to sixty appliances per household. The 1980s do not indicate any reversal in this trend—Americans' use of appliances will probably continue to escalate. This increase cannot continue forever. We must cut back sometime.

WHAT IS YOUR POTENTIAL WATT QUOTIENT (PWQ)?

Using the list of electrical appliances (see table), calculate your Potential Wattage Quotient (PWQ). This is done by multiplying the number of each appliance you have by the typical wattage for that appliance. If you use other electrical appliances not listed, contact your electric company and they will furnish you the average wattage used by a particular appliance. Then you should total the number of watts for all the appliances. Divide the total wattage by the number of people living in your house to calculate your Potential Wattage Quotient (PWQ).

Have students ask their parents to show them their last month's electric bill. Electric companies show the amount of electricity used in kilowatt hours. Kilo means one thousand; if your bill reads 862 kilowatt hours used (see Figure 1), you used 862,000 watts that month. How many kilowatt hours (or watts) did you and your family use last month?

---

*Alfred De Vito, "Watts Down is What's Up," *Science Activities*, Jan./Feb., 1976, Vol. 13, No. 1, pp. 14–15.

FIGURE 1

PUBLIC SERVICE				balance on last bill	payment applied	charges/ adjustments	new balance	amount billed	code

rate	reading dates present previous	days	meter readings present previous	multiplier	kilowatt hours used	kw demand	net charges	amount billed	
R06007.10 0609		31	97428 96566		862		41.17	41.17	
					SALES TAX		1.65	1.65	

2245 INDIAN TR DR
service address
120-164-1984
account number

BILLING DATE   JUL 16

DUE BY AUG 06   42.82

Your Potential Wattage Quotient (PWQ) only represents potential wattage, not actual wattage used. Your electric bill shows the actual wattage (expressed in kilowatt hours) used. A kilowatt is a unit of power, whereas a kilowatt hour is a unit of energy. Calculate your Actual Wattage Quotient (AWQ). This can be done by dividing the wattage (change kilowatts to watts) by the number of people living in your home. How does your PWQ compare to your AWQ?

Some More Energy Questions

How many bulbs do you have available for lighting in your room? Your home? What is the total potential wattage of all the bulbs used in your house?

How many electrical appliances are in your home?

How can you reduce your Actual Wattage Quotient? Name all the ways you can think of to reduce the amount of wattage used by you and your family.

Energy Consumption

Over the past twenty years our energy demand has been growing at a rate of 5% per year. At this rate, in the next twenty years, we will double our requirements for energy. This increase in energy demand would require further development of all the major energy sources such as oil, coal, oil shale, and nuclear power. The consequences of such a demand are difficult to foresee, but past experience has shown us that this demand would bring about undesirable consequences.

DID YOU KNOW?

In a single day, the United States uses approximately twenty million barrels of oil.

COMPUTE YOUR PWQ

Appliance	Typical wattage	How many of each appliance do you have?	Your wattage
Air conditioner	1,100 each		
Attic fan	400		
Central air conditioning	5,000		
Automatic toaster	1,200		
Automatic washer	700		
Broiler	1,000		
Built-in kitchen ventilating fan	400		
Clothes dryer	4,500		
Coffee maker	1,000		
Egg cooker	600		
Deep fryer	1,320		
Dehumidifier	350		
Dishwasher-disposer	1,500		
Dry iron or steam iron	1,000		
Electric blankets	200		
Electric clock	2		
Freezer	350		
Fluorescent lights (each tube)	15–40		
Garbage disposal	900		
Griddle	1,000		
Hair dryer	100		
Heat or sun lamp	300		
Hot plate	1,500		
Ironer	1,650		
Lamps, each bulb	40–150		
Mechanism for fuel-fired heating plant	800		
Mixer	100		
Oil burner	250		
Portable fan	100		
Portable heater	1,650		
Radio	100		
Ranges, electric	8,000		
Refrigerator	250		
Rotisserie	1,380		
Roaster	1,380		
Sandwich grill	1,320		
Stereo, hi-fi	300		
TV, black and white	350		
Vacuum cleaner	300		
Ventilating fan	400		
Waffle iron	1,320		
Waste disposal	500		
Water heater	2,500		
Water pump	700		

Total wattage

We have several choices. One, we could import as much oil as we can afford and wait and see which runs out first, the oil or our dollars. Two, we could develop alternatives to our common energy sources. This we will be forced to do. However, this development takes time and is expensive. Concomitantly, we are not always fully aware of the political, economic, or environmental impact of "new" energy forms. Three, we could reduce demand through energy conservation. This is an immediate choice since we now waste over half the energy we use.

TOTAL U.S.A. ENERGY CONSUMPTION*

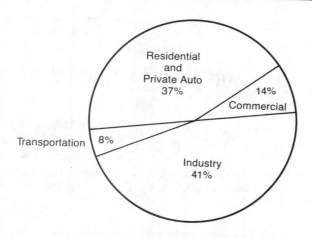

*Source: Park Project on Energy Interpretation, "Energy," National Recreation and Park Association, Arlington, VA, pg. 10.

HOW WE USE THE 37% THAT WE ALL CONSUME*

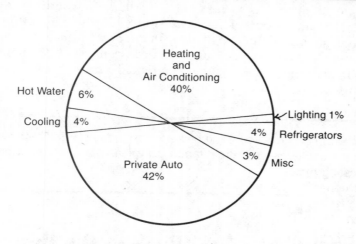

*Park Project on Energy Interpretation, "Energy," National Recreation and Park Association, Arlington, VA., pg. 10.

Everyone understands that you need monetary investments and time to produce oil and gas. The same is true in effecting conservation. Given past experiences and the current trend, we need to remember that developing energy conservation takes time and money. With oil supplies constricted for the foreseeable future and fuel prices on the rise, United States consumers may find that they simply have no choice but to use less energy and to use it more wisely.

DID YOU KNOW?

Through conservation techniques Americans can reduce their energy use by as much as 40% within a few years. This would be sufficient to eliminate the burden that imported oil puts on the United States economy and national security, with some savings left over.

What Is the Best and the Cheapest Energy?

The cheapest source of energy is conservation, or increased energy efficiency. The best energy is the energy you never use. Vehicles, equipment, buildings and appliances must do their work with a minimum of waste. The payoff of this kind of conservation could be an energy saving equivalent to several million barrels of oil a day by 1990, with little or no loss of economic growth.

DID YOU KNOW?

Transportation uses 55% of all the petroleum used in the United States. This amounts to roughly ten million barrels of oil a day or more than all the oil the United States imports.

Up to now, most American conservation efforts have been relatively easy housekeeping adjustments such as lowering thermostat settings, lights out at night, better maintenance of furnaces, and so forth. This effort has been considerable and has resulted in a saving of energy. Business, by contrast, has not done as well as they might have. Most companies have been unwilling to invest in energy-saving schemes that don't yield a substantial return in a short period of time.

DID YOU KNOW?

Americans will import twelve million barrels of oil a day by 1985—that will be 50% above the current level—and their import bill will amount to ten million dollars an hour at existing prices.

Should We Go "Back to the Basics" in Energy?

The answer to the question "How can we conserve energy?" would appear to be a simple one. Simply use less energy. But, nothing is really that simple.

One example of "back to the basics in energy" is the suggestion that we return to the horse and buggy days and stop using so much gasoline. Yes, we undoubtedly would save gasoline, but would we solve our problem? Think of a fictitious town somewhere in the midwest with a population of about half a million people. Suppose we assume that every three people in that town need a horse for transportation or general hauling work. This would mean we would need approximately 150,000 horses. These animals would need to be fed. To provide pasture and fodder for these horses would require nearly a half million acres of hay land and pasture, not to mention the acreage needed for growing oats.

If we wish to substitute horses for gasoline driven forms of transportation, we would have to ask questions such as:

Where are the horses?

Where is the crop production to feed them?

Where are the pastures?

Who would clean up the messy streets?

Could your town support all the horses it would need?

WHAT ARE OUR ENERGY CHOICES?

A lot of energy can be conserved it we alter existing habits. The following "less" and "more" list provides you with choices you can make to conserve energy now.

Less use of inefficient electrical appliances

More dependence on our own physical abilities to do work

Less frequent joy riding in our gasoline driven vehicles; also fewer solo passenger transportation runs to and from one place to another

More use of our public mass transportation vehicles

Less use of artificial climate apparatus such as home air conditioners and heaters

More use of natural insulators such as shrubs and trees, and use of more man-made insulation such as fiberglass

Less concern for individual comforts

More concern for the impact of your life style on the energy budget of the entire world.

A lot of energy can be conserved if, when you are faced with choices, you select the choice that involves the least amount of energy. For example, you could eat a cooked carrot or a raw carrot. One preparation uses much energy, the other little energy. Knowledge of the genealogy, or family energy tree, that a certain product possesses makes conservation decisions easier.

The cardinal rule to making wise, energy-saving decisions is to choose to stay as close to the natural state of the item as possible. Energy steps are required each time the original or basic item is changed to make a product. Picking the fig yourself and eating it uses less energy than buying figs in the store. Trucking, processing, baking and packaging the fig cookie takes a lot of energy. Thus, the conservation problem is compounded each time you increase the requirements for greater and greater refinement of a basic item.

Stay basic. When possible, select the natural state over the refined state. This reduces energy usage.

SUCCESSIVE STEPS OF A FRESH FIG TO A FIG COOKIE

DID YOU KNOW?

The energy from one gallon of gasoline is equivalent to 125,000 Btu's, 36,162 kWh, 10 pounds of coal, or 121 cubic feet of natural gas.

A human being requires 341 Btu's (or .1 kWh) of energy for one hour of normal activity. In one year, a human being would burn approximately 770 kWh or 21.1 gallons of gasoline. How many miles can your family car be driven on 21.1 gallons of gasoline? Would you trade one year's worth of your normal activity for what you might accomplish by being driven the distance 21.1 gallons of gasoline would take you? Is the gasoline engine as efficient as your body?

WHAT WOULD YOU BE WILLING TO TRADE?

Your body requires 0.1 kWh for one hour of normal activity. Check yes or no if you would make one of the trades listed in the following chart.

	Yes	No
Would you trade three hours and twenty minutes of your normal activity for one hour of viewing color TV (which uses .33 kWh an hour)?		
Would you trade two hours and thirty minutes of your normal activity for one hour of viewing black and white TV (which uses .25 kWh an hour)?		
Would you trade one hour of your normal activity for one hour of listening to a radio or stereo (which uses .1 kWh an hour)?		

Is There Any Such Thing As Easy Energy?

When there is a lot of something and it is relatively inexpensive, no one shows much concern. There is an old saying that goes, "What you get for nothing probably isn't any good." When energy cost almost next to nothing we knew it was good, but we squandered it as if it would never end. There is a direct correlation between ease of accessibility to energy sources and amount of energy usage. While many home ice cream makers are sold to people who are convinced that homemade, hand-cranked ice cream tastes better than ice cream purchased in grocery stores, much less ice cream would be consumed if we all had to crank our own. This is true of many of our daily requirements. Would you bathe as regularly if you had to chop the wood, start the fire and tend it to heat the water? Would you go to the movies if you had to walk five miles each way, crank the film, and heat the popcorn over a fire?

When a fish wants to get somewhere, it swims there. A bird flies to get somewhere. A rabbit hops to its destination. And, a turtle trudges slowly from one place to another. Humans have cleverly avoided walking. Whenever possible, we substitute some other form of locomotion for walking.

THE RANGE OF ENERGY INVOLVEMENT REQUIREMENTS BY HUMANS

HOW DO APPLIANCES COMPARE?

Compare the power used by different household appliances listed in *Watts Down is What's Up*. Divide the list of household appliances into three categories: absolutely essential, partially essential, and non-essential appliances. Does your list agree with one your neighbor may have assembled? Why? Or, why not?

TRACE YOUR ENERGY STEP

Record all those occasions during the day that you have a demand for energy. From the moment you wake up, notice when the bedroom light, TV, radio, oven, heater, electric toothbrush, bath (water heater needed), and all other energy users are on. Using the table of wattage of some common electrical appliances, (p. 101) add up your total for the day. You can find out how much energy you used for each appliance by multiplying the wattage rating by the length of time (in hours) it is used. Cut your daily total wattage in half. What effect would this action have on you? How might you cut down still further on your energy consumption?

DID YOU KNOW?

During the nineteenth century, Nantucket whalers insulated their homes from cold Atlantic winds by stuffing such things as seaweed, horse hair, and corncobs into wall cavities.

Insulation

When energy was cheap, home insulation was an added luxury. With today's rising energy costs, insulation has become a necessity. Insulation saves energy. It is estimated that approximately 90% of the homes in this country are under-insulated.

There are several things to consider if you are concerned with home heat loss. First, determine if there is any insulation in the home. Any material that reduces the passage of heat is insulation. Insulation does not heat or cool. What it does is restrict the flow of heat out of the places you are trying to keep warm, or into areas you are trying to keep cool. Any thermal insulating material that is dry, fire resistant, economical, easy to install, and, most importantly, contains numerous small cells to trap air and resist heat conduction, is a good insulating material.

DID YOU KNOW?

A technique using thermography (heat pictures) has been developed whereby heat loss can be precisely determined. Photographs of houses can be taken by a rapid scanning thermograph. The thermograph converts image data into numerical data which can be processed by computer. This is the same general approach used in detecting earth resources by satellite.

Usually attic insulation is exposed and readily visible. Wall insulation is hidden, so its presence is more difficult to determine. One way to check for it is to remove the cover from an electrical outlet on an outside wall. (*Caution:* Make sure that the electricity is turned OFF to the outlet.) Remove a small amount of wallboard or plaster around the outlet. Do not remove more plaster than will be covered by the outlet cover when it is replaced. You can visually check for wall insulation with a flashlight. Next, check how much heat can come in or go out through the windows. Glass is a poor insulator. A great deal of heat can be lost through the harmless-looking cracks around doors, windows, and foundation joints. Wherever two materials butt one against another, a joint or crack is present. This should be weatherstripped and/or caulked.

What Are R–Values?

The R stands for resistance to heat transfer. Thus, the R–value of insulation is simply a measure of how well a material resists or retards the flow of heat. Different materials have different R–values. The higher R–value, the greater the material's resistance.

R–VALUES FOR VARYING THICKNESS OF FIBERGLASS

Batts and Blankets	R–value
3½ inches	R–11
6 inches	R–19
6½ inches	R–22
9½ inches	R–30
12 inches	R–38

R–VALUES OF VARIOUS TYPES OF INSULATION PER INCH OF THICKNESS

Type of Insulation	R—value
Loose fill fiberglass	3–3.3
Loose fill rock wool	3–3.3
Loose fill cellulose	3.7–4.0
Loose fill vermiculite	2.0–2.6
Blown cellulose	3.1–4.0
Blown fiberglass	2.8–3.8
Foam (urea formaldehyde)	4.1–5.0
Urethane board	6.7–8.0
Fiberglass board	4.0–4.3

DID YOU KNOW?

Interior thermal shutters can improve storm windows' efficiency by 50%.

The following map shows the recommended amounts of insulation for each area of the country, based on average heating and cooling needs. Determine the zone of the country in which you live and then note the amount of suggested insulation.

What is the R–value for the area in which you live? Find out if the insulation in your home and/or school meets or exceeds this R–value.

The R–value of the insulation is usually found on a label on the package. If the packages are not labeled, ask what the R–value is before you purchase them.

SUGGESTED INSULATION FOR HEATING AND COOLING

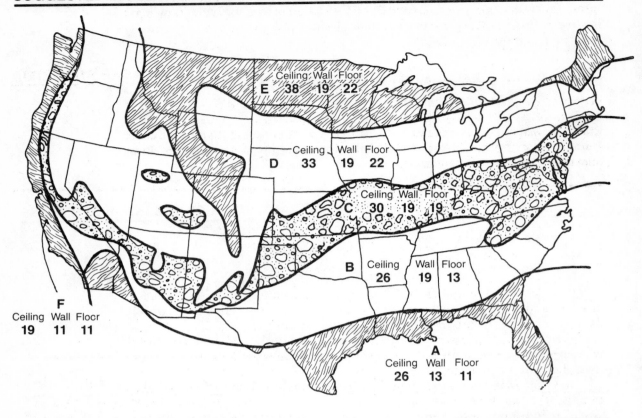

If You Live In Zone	Suggested Insulation		
	Ceiling	Walls	Floors
A	R–26	R–13	R–11
B	R–26	R–19	R–13
C	R–30	R–19	R–19
D	R–33	R–19	R–22
E	R–38	R–19	R–22
F	R–19	R–11	R–11

What Good Are Vapor Barriers?

Wet insulation loses its insulation value. Moisture in a home is desirable in the winter when the air is dry. And, it is undesirable in the summer when outdoor air is moisture laden. A vapor barrier is a material that acts as a seal to stop the transfer of moisture from one area to another. In a sense, it is an envelope that seals the interior of the house from the insulation. To be effective, the vapor barrier must be continuous and devoid of openings which might serve as passageways for moisture.

Some vapor barriers are attached directly to the insulation. In other instances, a separate vapor barrier must be installed before the insulation is put in. In either case, the vapor barriers should always be positioned between the heat source and the insulation.

THE ANSWER IS

In a house we use the vapor barrier to keep heat in and to keep the insulation dry and effective. When you dress for a winter sport that involves a lot of energy and during which you will work up perspiration, why don't you dress in a plastic barrier and cover it with several sweaters? What is the difference between insulating a house and a human? The answer is ventilation.

What Is Ventilation?

Ventilation is the movement of air through a confined area. It is the most efficient and least expensive way to prevent excessive humidity from causing damage to a building, particularly during the heating seasons.

Warm air has a greater capacity for holding moisture than cold air. Thus, when a room is ventilated, the transferred or escaping warm air has more vapor than the cold air replacing it.

In small areas such as kitchens and bathrooms, small fans should be used to expel the highly moisture-laden air. Larger areas, such as attics and crawl spaces, need continuous, summer and winter ventilation. This constant ventilation is best achieved by two opposing vents that permit air to flow in one vent and out the other.

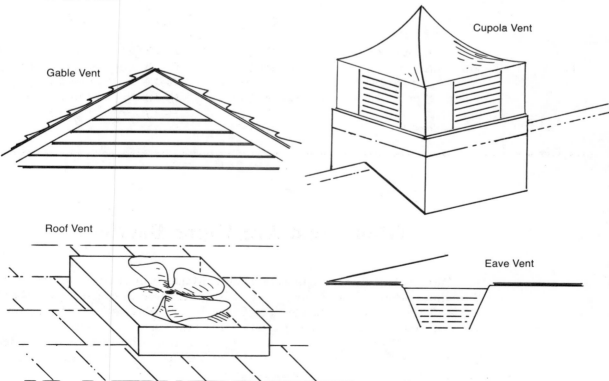

Gable Vent

Cupola Vent

Roof Vent

Eave Vent

Attic fans are excellent for venting attic air. These fans work so well that in areas where summers are mild, they are actually used in place of central air conditioning.

What Are the Various Types of Insulation?

There are five basic forms of insulation used to save energy in homes. These are batts, blankets, loose fill, foam, and rigid form.

Batts are usually made of fiberglass or rock wool. They come in 4- or 8-foot lengths and 15- or 23-inch widths. They are moisture-resistant and fire-resistant and are available with or without a vapor barrier attached. Although these precut lengths are easy to handle, when one piece butts up against another the continuity of vapor barrier and insulation is interrupted and an air passageway may be constructed.

Blankets are exactly like batts except that they come in a continuous roll and are cut to fit specific areas.

Loose fill has the advantage of being able to be poured or blown into desired insulation cavities. No vapor barrier is present. Before installing loose fill in an attic, a continuous plastic vapor barrier at least 2 mils thick should be placed over the flooring beams and flattened down between them.

Rigid insulation may be made of extruded polystyrene, urethane, or fiberglass. Although urethane and polystyrene are efficient because of their high R— value, they are both combustible and absolutely must be covered with ½-inch thick wallboard for fire safety.

Fiberglass boards are also effective as insulation and are commonly used for basement walls.

MAKE A DRAFT GAUGE

To locate sources of drafts that need insulating you will need: a metal clothes hanger, a plastic sandwich bag (or a light piece of tissue paper) a pair of scissors, and two clothes pins. Cut the sandwich bag down each side and wrap one end over the cross-bar of the clothes hanger. Use the clothes pins to fasten the bag to the bar.

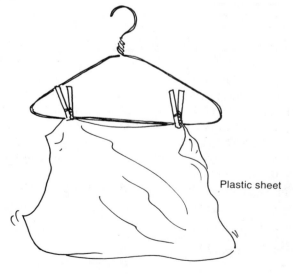

Plastic sheet

To check for drafts around a window, hold the apparatus by the handle of the hanger and place the edge with the plastic bag clipped to it close to the edge of the frame. If any breeze is coming in, the movement of the plastic will show you where to caulk the window frame. Try the draft gauge on other suspected draft areas.

Find the Best Insulator

Collect five small identical jars with lids. Baby food jars will do nicely. Also collect four half-gallon milk cartons. Fill each of the milk cartons with one of the insulations you have chosen—the insulation can be shredded newspaper, fiberglass, sawdust, cloth, etc. Then place one jar in each milk carton. One jar should be placed in the milk carton without insulation to serve as the control. Place one ice cube (approximately the same size) in each jar and cap it. Push the jar down into the insulation so that it is completely surrounded by insulation. No insulation is added to the control jar. When the uninsulated ice cube has melted, check the others. Which ice cube melted the most slowly? Which insulation turned out to be the best?

ENERGY SAVER'S CHECKLIST

Yes   **No**

- Do you have combination doors?
- Do you have combination windows?
- Are plastic sheets stapled to and covering windows?
- Is there caulking around:
    fans and air conditioners
    window frames
    door frames
    chimney
    corners formed by siding
    the foundation where it meets the house siding?
- Are you using window shutters and awnings?
- Do you have weatherstripping on windows and doors?
- Have you insulated all the heating ducts that run through unheated areas?
- Do you replace heating air filters once a month during the heating season?
- Do you vacuum air ducts and registers every month?
- Are you releasing or "bleeding off" air from upper floor radiators when hot water systems are used?
- Have you lowered the temperature setting on your hot water heater? Set it at 120 degrees F (140 degrees F for dishwasher use).
- Is your hot water tank insulated?
- Do you periodically (six months to a year) drain one gallon of water out of the tank to remove sediments that may accumulate and retard efficiency of the heater?
- Do you use a heat pump? In a climate where both heating and air conditioners are needed, the heat pump is an energy-saving device. The heat pump does not create heat, it simply moves heat. In the winter, it takes heat from the outside air and brings it inside. In summer it reverses the process.
- Have you turned down the house thermometer?

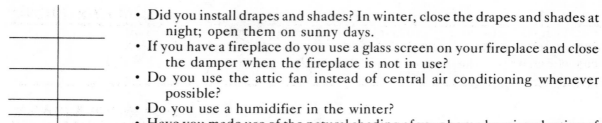

- Did you install drapes and shades? In winter, close the drapes and shades at night; open them on sunny days.
- If you have a fireplace do you use a glass screen on your fireplace and close the damper when the fireplace is not in use?
- Do you use the attic fan instead of central air conditioning whenever possible?
- Do you use a humidifier in the winter?
- Have you made use of the natural shading of your home by wise planting of trees and shrubs?

DID YOU KNOW?

Household furnaces are typically said to have an efficiency of about 0.6—this means that 60% of the heat from the combustion of the fuel is delivered as useful heat to the house and 40% is lost up the chimney.

If you use an electric clothes dryer, vent the exhaust into the house to add heat and humidity to the indoor climate during the winter months. This process usually involves the installation of a special, indoor-venting device. If excessive condensation and dust become problems, this action may have to be discontinued.

When using air conditioners, set the thermostat at 78 degrees Fahrenheit. Close all doors (including closet doors), windows, and unused rooms while the air conditioner is running. When the house is going to be vacant for long periods of the day, turn down the air conditioner before leaving. Also, close drapes and window shades when operating your air conditioner, but be careful not to block or interfere with the flow of cool air. Also check air-conditioner filters at least once a month. Wash and replace them as needed.

DID YOU KNOW?

As much as 10% of a home's heat can filter through the holes in electrical outlets.

The following list gives you the R— values of exterior household materials that save energy.

Material	R—value
airspace between materials, 1½ inch	0.97
facing brick, per inch	0.11
concrete, poured, per inch	0.08
concrete block, 8 inch	1.04
lightweight concrete blocks, 8 inch	2.18
plywood, per inch	1.25
plywood, ⅜ inch	0.47
plywood, ½ inch	0.62
hardboard, medium density, per inch	1.37
gypsum wallboard, ½ inch	0.45
lumber (softwood), per inch	1.25
asphalt shingles	0.44
wood shingles	0.94
styrofoam, 1 inch	5.00

DID YOU KNOW?

Reducing the thermostat reading by 5 degrees could save you as much as 15 to 25% on your fuel bill.

WHAT IS THE ANSWER?

Some Americans think they have discovered a solution to the high cost of the home heating problem by turning to wood-burning stoves. Perhaps they have, but experts now warn us that smoke from burning wood contains toxic emissions as dangerous as those from oil and gas.

In the town of Portland, Oregon, more than one-third of the soot and ash on a winter's day comes from the burning wood used to heat homes. Wood pollution contains potent carcinogens similar to those found in cigarette smoke. You can minimize the release of toxic gases by burning small amounts of dry wood at a time, providing sufficient air, and keeping the temperatures high. Toxic gases should be burned within the stove. This can be accomplished by installing a baffle plate in the stove for good air circulation.

CHECK YOUR HOME FOR THESE INSULATION ENERGY SAVERS

	Yes	No

- Insulation of:
  - attics
  - walls
  - floors
  - ceilings
  - between the house and
  - unheated garage
  - hot water pipes
  - heating ducts
- Use of vapor barriers
- Installation of an attic fan to replace air conditioning
- Installation of storm windows
- Doors, windows and other openings caulked and weather-stripped
- Use of heavy window shades and drapes that can be opened and closed as needed
- Use of shade trees, bushes, or awnings to keep heat out during the summer months

DOES INSULATION HELP SAVE ENERGY?

Add a ½ cup of hot water to two, equal-sized plastic foam cups. Record the temperature of the liquid in each container. Take one cup and insert it inside three nested cups. Record the temperature in each container every five minutes for a period of one half hour. Does the increased insulation affect the original temperature of the hot water?

Repeat this experiment, but this time cover each cup with a piece of plastic foam with a hole in it—the hole allows the thermometer to stick out. Does covering the cups have an effect on the temperature of the water?

DID YOU KNOW?

Trees planted around a residential home can provide a 40% reduction in heating costs in the winter, and their shade provides even greater benefits in the summer. Trees can also be harvested for a profit.

DID YOU KNOW?

Foresters are attempting to regulate temperature and humidity and to lower pollution by altering the types and location of vegetation around buildings and open space.

Kitchen Energy Savers

The two most used rooms in a house are the kitchen and the bathroom. The kitchen is an excellent place to start conserving energy. Here are some conservation suggestions:

• If you have a gas stove, make sure the pilot light is burning efficiently. A yellow flame indicates an adjustment is needed.
• If you need to purchase a new gas oven or range, purchase one with an

electronic ignition system instead of pilot lights. Up to 47% of the gas used by an oven or stove can be saved by the use of an automatic ignition system.

- Keep the stove reflectors and burners clean and bright. They will reflect the heat better and thus use less energy.
- If you use an electric range, turn off the burner several minutes before the desired cooking time is reached. The residual heat in the element will stay hot long enough to finish the cooking without using more electricity.
- Don't use oversized pots or pans to cook things. If the pot or pan hangs over the edge of the heating element, heat is wasted heating up the excess metal in the oversized pan.
- Cook as many foods in the oven as possible at one time. (You may want to plan meals you can prepare and freeze, or that will keep for a long time.) Make the oven heat serve more than one cooking task.
- Never use the large oven area to heat small meals, use a smaller appliance such as an electric pan or toaster oven. Less energy is used by the smaller unit.
- Don't open the oven door every few minutes to check food. Use a timer. A great deal of heat escapes every time you open the oven door.
- Never boil water in an open container. Water will boil faster and use less energy when boiled in a kettle or covered container.
- Never heat any more water than you need to accomplish a specific cooking requirement. Heated water wasted is energy wasted.
- When possible, use cold water for tasks such as cleaning the sink, rinsing out a bottle, and so forth. Unheated water uses less energy than heated water.
- Repair leaky faucets. One drop per second can waste as much as 175 gallons of hot or cold water per month.
- Only use your dishwasher when you have a full load of dishes. You use as much water for a less than full load as you use for a full load.
- Scrape dishes before loading your dishwashing machine. This allows you to skip the rinse.
- Don't use the "rinse hold" function on your machine. The rinse hold uses 3 to 7 gallons of hot water per wash.
- Let your dishes air dry. Air drying saves on the energy required to heat dry dishes. In winter the moisture from the drying dishes contributes to the desired indoor humidity level.
- Use a minimum amount of water for your purposes. Unnecessary heated water wastes energy.
- Turning lights off does save energy. It can also waste energy if some precautions are not observed. Frequent switching on and off of light bulbs can substantially shorten the life of a bulb. If you plan to leave a room lit by incandescent bulbs for more than three minutes, turn off the lights. If you'll be gone less than three minutes, leave them on. If the room has fluorescent bulbs, only turn the lights off if you plan to be gone for more than fifteen minutes. More energy is used to turn a flourescent light off and on within a fifteen minute period than the amount of energy consumed by the bulb within a fifteen minute period.
- Use aluminum and copper cooking utensils instead of stainless steel utensils. Aluminum and copper are better conductors of heat. Stainless steel utensils with aluminum or copper bottoms are preferred for improved spreading of heat. Cast iron utensils heat up slowly and should only be used when you intend to do a great deal of cooking. For short heat requirements cast iron utensils waste too much energy.

- When replacing your current dishwasher, purchase one with an automatic turn-off feature. This automatically turns the dishwasher off after the rinse cycle. This can save up to one-third of the total dishwashing energy costs.
- Use energy-intensive appliances—clothes washers and dryers, dishwashers, and electric ovens—at times when the nation's electrical system is not at a peak load. Peak load time is usually during late afternoon and the early evening hours. Thus, a more efficient time to use energy-intensive appliances would be in the early morning or late evening hours. Limit the times you use these energy-intensive appliances. Maximize the work done by the minimum amount of energy.
- Make sure your refrigerator door seals are airtight. Slip a piece of paper between the refrigerator door and the door frame. If you can pull the paper out easily the seal may need replacing or you may need a new door latch.
- Don't keep your refrigerator colder than it needs to be. Thirty-eight to forty degrees is recommended for the fresh food compartment; 5 degrees for the freezer section, and 0 degrees for the freezer.

DID YOU KNOW?

If every dishwasher in the country was used one less time per week, we would save the equivalent of about 9,000 barrels of oil a day.

WHAT IS EER?

EER (energy efficiency ratio) is a convenient number that indicates the efficiency of an appliance such as an air conditioner or refrigerator. It is a measure of the amount of Btu's achieved relative to the amount of electricity used. EER is determined by dividing the Btu per hour rating of an appliance by the watts (power) it uses. For example, one 8,000 Btu air-conditioner consumes 1,000 watts.

$$\text{The EER is: } \frac{8{,}000 \text{ Btu/hr}}{1{,}000 \text{ watts}} = 8.0 \text{ Btu/watt-hour}$$

Another model of the same capacity (8,000 Btu/hour) might consume only 800 watts. It is more efficient since it accomplishes the same amount of cooling with less power consumption.

$$\text{The EER is: } \frac{8{,}000 \text{ Btu/hr}}{800 \text{ watts}} = 10 \text{ Btu/watt-hour}$$

The higher the EER of an appliance, the more efficient the unit and the less power it will consume. Since it is more efficient, the unit may cost a little more, but will probably be a better buy in the long run when you consider the energy savings it delivers.

Visit an appliance store and obtain sample EER tags of appliances within a category, like dishwashers or air conditioners. Next, rank the tags according to their EER (from highest to lowest). Is the highest EER-rated appliance also the highest price? Which appliance would you buy?

Conduct a survey in your school or around your neighborhood. Have people tell you what they think EER means and how it is used.

FLUORESCENT OR INCANDESCENT—
WHICH GIVES THE MOST LIGHT?

Does a fluorescent lamp give off more light than an incandescent lamp of the same wattage? You will need a light meter to measure any differences. A light meter can be borrowed from a photographer or a camera enthusiast. Choose a light source and record the amount of light it gives off when you stand 1 meter away from it. Next, measure and record the amount of light it gives off 2, 3, and 4 meters away from it. Repeat this experiment with a few other light sources. Graph your results. How do they compare? What are your recommendations?

Using the same light meter, can you determine how much brighter a 100-watt incandescent lamp is than a 60-watt incandescent lamp?

Other Places To Save Energy

- Wash your face and hands in cold water instead of hot water. Whenever you have a choice between using warm or hot water for washing your face or hands, washing clothes, or for any other process, reach for the cold water tap first.
- Take showers instead of baths.

DID YOU KNOW?

Money goes down the drain. One drip per second from a leaky hot-water faucet or showerhead sends about 175 gallons a month down the drain.

HOW MUCH DOES A DRIP COST?

Materials Needed

8 ounce measuring cup

paper

pencil

clock or timer

Adjust a sink tap (cold water) to produce a steady drip, drip, drip. Catch ten minutes worth of drips in the glass. The following example shows you how to figure out the amount of energy being wasted by that drip. The example is based on 3 ounces of water in the glass.

1. Multiply the number of ounces by 6 to give the number per hour and then multiply that figure by 24 to give the total per day.

    Example: 3 ounces/10 minutes times 6 (min/hr) times
    24 hours/day = 432 ounces/day

2. To find the number of gallons that would be wasted per year, multiply the ounces per day figure by 365 days and divide by 160 ounces in a gallon.

    Example: 432 ounces/day × 365 days/year ÷
    160 ounces/gallon = 985.5 gallons/year

3. To find the number of Btus used to heat the water you must first know the temperature of the tap water before and after it is heated. Assume the tap water temperature is 40 degrees Fahrenheit when it enters the hot-water heater, and 140 degrees Fahrenheit after it is heated. Energy must heat the water 100 degrees Fahrenheit. To find the Btu required, multiply the number of gallons per year by 10 pounds (weight of a gallon of water) and by the 100 degrees Fahrenheit difference.

Example: 985.5 gallons/year × 10 pounds
× 100 degrees Fahrenheit = 985,500 Btu

4. Once you know the Btu, you can find out how many gallons of oil, cubic feet of gas or kilowatt hours of electricity it took to heat the wasted water.

Divide the following set of figures into your answer of 985,500 Btu. This will tell you how much fuel or electricity is used for heating the water.

One gallon of heating oil = 138,800 Btu
One cubic foot of natural gas = 1,031 Btu
One kilowatt hour electricity = 3,414 Btu
Example:
985,500 Btu ÷ 138,800 = 7.1 gallons of oil
985,500 Btu ÷ 1,031 = 955.9 cubic feet of gas
985,500 Btu ÷ 3,413 = 288.7 kilowatt hours

5. How much is this drip actually costing you? Multiply the appropriate figure from Step 4 by what you are paying for your method of heating. For example, if you are paying five cents per kilowatt hours for electricity, the cost in the example would be:

288.7 kWh × $0.05 = $14.44

How much money is going down the drain in your home or school? Find out.

DID YOU KNOW?

It takes about 30 gallons of water to fill the average tub. A shower, with a flow of 4 gallons a minute, uses only 20 gallons in a five-minute period.

• Use one large light bulb to substitute for several smaller ones wherever you can.
• Don't run your appliances unneccesarily.
• Turn off all radios, TVs or record players when they are not in use.

DID YOU KNOW?

One 150-watt incandescent bulb puts out 2,880 lumens (a measure of light intensity). Two 75-watt incandescent bulbs produce only 2,380 lumens.

DOES IT MAKE SENSE?

Does it make sense to use an incandescent light bulb when you could use a fluorescent tube? A 40-watt deluxe warm white fluorescent tube generates 2,150 lumens. A 100-watt incandescent light bulb glows with an intensity of only 1,750 lumens.

A new light bulb has recently been introduced to the public. This bulb burns for about five years, uses one-third the energy of existing bulbs and can save about twenty dollars over its life in lower electric bills. The bulb burns no more than 50 watts of electricity while shedding as much light as a 150-watt incandescent bulb.

HOW MUCH ELECTRICAL ENERGY DOES YOUR FAMILY USE IN ONE WEEK?

Make an appliance tally tag for each appliance in your house. This tag will list the name of the appliance, an approximate requirement in kilowatt hours for that appliance, and date and time recording columns. Your tally tag might look like this.

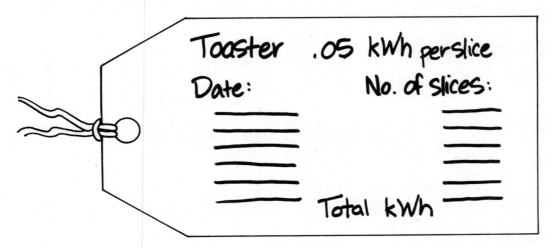

Each time you use each appliance, record the length of time that particular appliance is in use. Do this for all appliances used for a period of one week. To keep an accurate record, ask all members of your family to record the time they use each appliance on the tally tags.

The following table provides some estimates of kWh for most appliances. It is not necessary to record the hourly rate for the refrigerator, freezer, or range. These are calculated for you on a per day or per meal basis. Thus, they can be calculated and included in the total weekly figure.

At the end of the week, add up the total kWh. Check with your electrical service company for the kWh rate. To obtain the total cost of electricity for a one week period, multiply your electric company's rate by the total kWh used.

Compare this electric bill to last year's electric bill. Remember, this will be a monthly total rather than a weekly total. Has the price increased? By what percentage?

KILOWATT HOURS FOR COMMONLY USED HOUSEHOLD APPLIANCES

Kitchen Appliances		Bathroom	
Refrigerator, frostless, 16 cubic feet	5 kWh a day	Hair dryer	.33 kWh an hour
		Razor	.001 kWh a shave

Freezer, 15 cubic feet	5 kWh a day	Toothbrush	.001 kWh per brushing
Range	1.5 kWh per meal	Bath	1 kWh per bath
Oven	.50 kWh an hour	Shower	.50 kWh per shower
Dishwasher	1 kWh a load		
Garbage disposal	.01 kWh a load		

Laundry Room

Bedroom

Washing machine	33 kWh a load	Electric blanket,	
Clothes dryer	3 kWh a load	king	1 kWh a night
Iron	1 kWh an hour	queen	.75 kWh a night
Water heater	12 kWh a day	twin	.50 kWh a night
		Portable heater	1.5 kWh an hour
		Clock	1.5 kWh a month
		Sewing machine	.02 kWh an hour

Living Room

Kitchen Utensils

Color television	.33 kWh an hour	Toaster	.05 kWh a slice
Black & white TV	.25 kWh an hour	Waffle iron	.25 kWh an hour
Radio-phonograph	.10 kWh an hour	Coffee maker	.25 kWh a pot
Vacuum cleaner	.67 kWh an hour	keep hot	.50 kWh an hour
100-watt bulb	.10 kWh an hour	Blender	.02 kWh per operation
		Frying pan	1 kWh an hour
		Mixer	.05 kWh per operation

FIND THE ENERGY WASTERS

Think of all the ways energy is wasted. Walk from room to room in your house and write down all the energy wasters in each of them. Next, write down what should be done to save energy in each instance you've just recorded. Remember to include things like open windows when the heat is on, electric blankets warming up, refrigerator doors held open, unnecessary lights on.

Discuss how you will turn energy wasters into energy savers.

Transportation

DID YOU KNOW?

If all the automobiles in the United States were run simultaneously, their electrical generating capacity would equal about fifty times that of all our central power generating stations.

DID YOU KNOW?

The United States accounts for 6% of the world's population and 46% of the world's cars.

Automobiles occupy a special place in the hearts of Americans. So much so that Americans have been accused of having a "love affair" with their automobiles. We have grown to love them bigger, better, faster, stronger, sleeker, and so forth. The gasoline that provided the energy to power these machines seemed to be almost endless in supply. But, our appetite for bigger cars, more cars (two or more per family), and less fuel-efficient cars caught up with the energy supply. Transportation of humans and the goods that humans need accounts for one quarter of our annual energy consumption. This energy consumption is inefficient and primarily uses the fossil fuel, oil.

An automobile's fuel economy is determined by its weight, size, engine type and size, manual transmission, low axle ratio, and maintenance. Small lightweight cars are more economical to operate than full size, heavy cars. In general, a 5,000 pound car uses twice as much fuel as a 2,500 pound car (in local driving).

Optional features like air-conditioning, automatic transmission, and power steering require more energy, which is derived from burning gasoline. Other accessories such as power brakes, seats, and windows use less energy, but a car without them is more energy efficient.

DID YOU KNOW?

An automobile's air-conditioner lowers fuel economy 10 to 20% when used in stop-and-go traffic.

DOES IT MAKE SENSE?

Does it make sense to purchase a small, economical car to conserve energy and then to add features like air-conditioning, power brakes, power steering, power windows, and a power radio antenna?

How can you conserve energy in transportation?

• Drive only when it is absolutely necessary. Eliminate the numerous one item, one trip runs to the store. Think before you drive. Consolidate your buying. Shop once a week instead of daily. Handle more problems by phone or by mail instead of driving to the location and talking to the person.

DID YOU KNOW?

If every automobile took just one less ten mile trip a week, we could save 3½ to 4 billion gallons of gas a year.

• Shop with friends. One car per person is wasteful.
• Share your transportation. Create carpools. About one-third of all private automobile mileage is for commuting to and from work. Many employers provide vans and assign drivers to them. The van completes a regular daily route to and from work. The riders pay only for the gas. The driver travels free. This compensates him for assuming the driving responsibility. Remember, cool it. Pool it.
• Use alternative methods of transportation whenever possible. Use a motorcycle, a bicycle, skates, or, best of all, walk.
• When driving, drive energy efficiently. Drive at the 55 mph speed limit. The best fuel economy occurs at speeds of 30 to 40 mph with no stops or rapid speed changes.

DID YOU KNOW?

If you increase your automobile speed from 55 mph (the legal limit) to 65 mph, the automobile's resistance to the wind goes up by 40% and fuel economy drops. Most automobiles get about 20% more miles per gallon at 55 mph than they do at 70 mph.

• Don't jack rabbit your automobile. Accelerate and decelerate slowly and smoothly. Quick starts and stops waste gas. Keep the car moving at a smooth, steady pace. Reduce braking as much as possible by anticipating stops and gradually slowing down in preparation for them.

• Keep your automobile in good running order. Have your car tuned regularly. Keep your engine's oil, air, and gas filters clean. Use the recommended octane rating for your automobile. Keep your tires balanced and filled to the correct air pressure. Also, try to switch to radial tires. Radial tires outperform conventional tires. They last longer and provide 3 to 5% better mileage in the city. Radial tires can increase your car's highway mileage by 7 to 10%.

DID YOU KNOW?

A poorly-tuned car engine uses 3 to 9% more gasoline than a well-tuned car engine.

• Don't carry excess baggage in your car. Items like golf clubs, excess tires, and tools increase the total weight of the car, so that more energy is used to keep the car moving.

DOES IT MAKE SENSE?

Some automobile owners are storing a gasoline-filled can in their car trunk in reaction to the recent oil shortage. This is extremely dangerous. Drastic temperature changes in the car trunk can occur and cause the gasoline to expand and even rupture the gas container, allowing gas fumes to leak out. A spark from anything—a short in a taillight or a rear end collision—could cause an explosion.

CAUTION: One gallon of gasoline stored in the trunk of the car has the explosive power equal to fourteen sticks of dynamite.

DID YOU KNOW?

Oil, or petroleum, was known about for centuries before humans put it to use as a source of lighting and power. Petroleum oil was and is still used for its medicinal values. Often it was sold under the name of "snake oil."

Recycling

The story of energy is one of flow and change. Energy can exist as coal, oil, wood, or gas, but with usage it constantly changes form. Coal, oil, wood, and gas possess potential energy. When these materials are burned in a power plant furnace, they are changed to a kinetic form of energy—the heat energy of steam. This steam energy turns the turbine engine which turns an electric generator. The energy coming out of the generator is electric energy which performs a multitude of tasks. So energy is not really used up; instead, it just changes from one form to another. If this is so, how can there be an energy shortage? According to the first law of thermodynamics, which states "energy can be neither created or destroyed," there should never be a shortage. This law is accurate but is not totally comforting because it is followed by the second law of thermodynamics which states, "In any transformation of energy from one form to another, some of it becomes unavailable for further use." Or, described in other terms, this second law says that it is impossible to do something with the same energy that was used to do work in the first place. Once the energy involvement is accomplished, the energy to accomplish it has degraded to a lower level and less is available to do work. It becomes unavailable because it becomes heat and it is not possible to convert a given amount of heat energy back into another form. It is the second law that explains the energy crisis.

Whenever energy is used, it loses some of its quality. This phenomenon is known as the entropy law. An example of the entropy law is burning a kitchen match to heat up a thimble full of water. The temperature of a burning match is high. When the burning match is placed under the thimble, some of its heat raises the temperature of the water. The remainder of the heat from the burning match is lost to the air, making the air a bit warmer. Thus, no energy is lost in the process; it is simply spent in several directions. In the process of burning, energy is altered from high quality availability to low quality availability.

Renewable and Non-renewable Energy Resources

All energy resources belong to one of two groups—renewable or non-renewable resources. Renewable energy sources are those which can be replenished or are non-depletable. Water is replenishable—we use it and then we discard it. Water is returned to us to be reused over and over again in the hydrologic cycle. Fossil fuels such as coal, oil, and gas are non-renewable, even though some of these materials are currently being formed. The formation of the fossil fuels takes millions of years; and, if we continue to consume them at our present rate, we will use them up before those being formed are available for continued use. Fossil fuels are finite. No matter how skillful we become in the exploration of fossil fuels, the end of their availability is inevitable. This is not the case with renewable resources which are parts of cycles and are viewed as continuous energy sources. Some examples of renewable sources are solar energy, wind power, photosynthetic energy, and energy from waste.

Waste can be renewable when it becomes part of a cycle. Each year millions of tons of combustible waste materials are thrown away. The methods we are now using to dispose of this large volume of solid waste are creating serious environmental problems for many, many cities and communities.

The total energy potential from one year's worth of our organic solid waste is equivalent to 20 to 25% of the energy we currently derive from oil. Organic solid waste is an almost untapped source of energy. Although organic solid waste is available for use, it is not easily obtainable because it occurs in small amounts at numerous locations. Nevertheless, sufficiently large cattle-feeding lots, large sawmill operations, and municipal waste collections, if collected and processed, could account for approximately 200 million barrels of oil per year.

Waste consisting of organic and inorganic (glass, metal, plastics) matter can be processed in a variety of ways to extract energy. The organic waste can be shredded and dried, and the inorganic material can be removed for recycling. The organic material can then be reshredded and heated to 500 degrees Celsius in an oxygen-free atmosphere. One ton of organic refuse processed this way can produce one gallon of oil plus some accompanying gas.

Another technique that can be used to conserve energy is mixing shredded garbage with coal and burning it in power plants.

A more complicated energy derivation process is the method of converting organic waste to usable fuel by hydrogenation. In this process, waste is heated at high pressure and temperature in a carbon monoxide plus steam environment. In the process approximately 99% of the carbon content of the organic waste is converted to fuel oil. Waste processed in this manner would yield one and a quarter barrels of oil per ton of dry waste.

A simpler, more natural process of extracting energy from organic waste is to allow the natural process of anaerobic (oxygen free) digestion of organic material by microorganisms. This process produces methane gas. One ton of solid waste can generate about 10,000 cubic feet of methane gas. Methane gas is a clean, desirable fuel.

DOES IT MAKE SENSE?

Fossil fuel is used to make the nitrogen fertilizer used in planting, cultivating, and harvesting an annually repeated corn crop in the same location. If a crop of legumes is rotated with corn every two years, 25% less fertilizer is needed, and thus about 12% less energy is used over a three year period.

DID YOU KNOW?

The throwaway, lowly corncob is actually a little bundle of energy. It can fuel a fire nicely. Also, the corncob pith is used as an animal feed extender and the woody ring material makes an excellent abrasive. It is also used for explosives, fertilizer extenders, insulation materials, and can be molded into furniture parts. It can also be used as a soil conditioner.

Inorganic materials, such as metals, glass, and plastics, can be recycled. An excellent example is the recycling of automobiles. Giant shredders can rip an automobile into fist-size pieces of scrap in less than a minute. Shredded scrap passes over a magnetic drum which separates iron and steel from other metals and non-metallic materials.

DID YOU KNOW?

One ton of scrap saves 1½ tons of iron ore and ⅓ ton of coal/coke.

DID YOU KNOW?

Discarded aluminum cans are worth about $350.00 a ton. New metal for cans from used aluminum cans saves 95% of the energy needed to produce aluminum from ore. And aluminum can be recycled again and again.

DOES IT MAKE SENSE?

Some governmental policies discourage the recycling of metallic scrap. It costs about 2.6 times as much to ship scrap iron by rail as it costs to ship its competitor, iron ore. Tax incentives, such as depletion allowances, are available to the producers of iron ore, but no such incentive exists for recycling metallic scrap. This in essence discourages conservation efforts and encourages the continued mining of irreplaceable natural resources.

THE ANSWER IS

The gap between energy supply and demand can be reduced by using only what is essential. The energy education motto is, "Do more with less."
- Do more work yourself.
- Use less labor-saving devices.
- Do more walking.
- Use less private automobiles.
- Use recycling centers.
- Don't throw anything away. Reuse it.

OVER AND OVER—RECYCLING

- Find out about the recycling efforts in your community.
- What do businesses and industries in your area recycle?
- Begin a recycling campaign at your school. Try recycling: aluminum cans, tin cans, glass bottles, and newspapers.
- Have people bring in toys and games that they no longer want and recycle them by redistributing them.
- Demonstrate how paper is recycled by tearing up old newspapers into small strips, place in a bowl, add water, and mix with a hand mixer. Strain the water out of the paper by using a wire screen and let the recycled paper dry.
- Report examples of "doing more with less." Here are two examples to get you started.
  A single chip of silicon (one of the most abundant substances on earth) a tenth of an inch square can perform as many functions as one thousand separate electrical components.
  The telstar satellite, which weighs only one-tenth of a ton, outperforms seventy-five thousand tons of transatlantic cable.
- Find out what your local historical society is doing to recycle homes and businesses in your community. Is your school going to be recycled in the twenty-first century? Why or why not?
- Prepare recycling posters that emphasize the importance of recycling furniture and clothing, metal hangers, paper, newspapers, newspaper logs for fireplaces, using both sides of a piece of paper, using cloth towels instead of paper towels, reusing food containers, and growing your own garden.

DID YOU KNOW?

There are over twenty million junk automobiles in the United States today—the reusable metal in them is worth over one billion dollars.

DID YOU KNOW?

Each ton of recycled paper saves seventeen trees and the energy to process them.

Can We Build and Redesign for Energy Efficiency?

Most of the homes in the United States were constructed with little concern for energy efficiency. Taking energy-saving measures when building a new home often adds to the building costs, but these measures lead to long-term benefits such as increased comfort and lower operating costs. The long-term savings are usually two to three times more than the initial investment.

A number of energy factors should be considered when selecting a building site. For example, a house located on the southern slope of a hill derives maximum benefit from the sun. Homes built into a slope can profit from the soil surrounding the lower portions of the house. Soil can act as good insulation to cold, above-ground temperatures.

Trees adjacent to a potential building site should be given special consideration in the energy efficiency planning. Plan to locate the building site within a reasonable distance of wooded areas. Planting trees on all sides but the southern portion of the house is the most advantageous position. Trees and wooded areas provide a windbreak and help reduce heat loss. An open southern exposure utilizes incoming solar heat.

DO WINDOWS LET MORE THAN LIGHT IN AND OUT?

Which side of your classroom has the greatest outside exposure? Which compass direction does it face: north, south, east or west? How much of this side of your classroom is made of glass? Compare this glass area to the remaining wall of the same side. Feel the glass and the wall. Which material is the best insulator? Would you save energy if you had fewer windows?

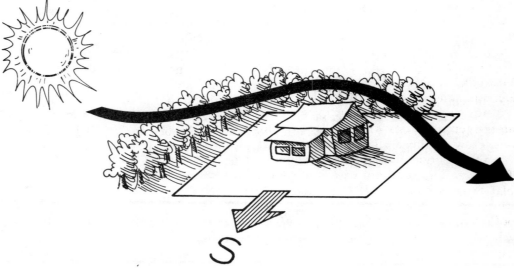

DO WINDOWS AFFECT HEAT GAIN AND LOSS?

Take two empty half-gallon milk cartons. Cut a window out of the center of one side of one of the cartons. Cover this cut out area with plastic wrap and tightly seal the plastic to the milk carton. On the side opposite the window, punch a small hole to allow a thermometer to be inserted. Wrap a rubber band tightly around the thermometer so that it does not fall through the hole. Punch a similar thermometer hole in the remaining milk carton. Staple the tops of both milk cartons closed.

Find a window in your house with a southern exposure. Position the two milk cartons in direct sunlight. Use a compass to determine due north from that window. Place the carton with the window in the sunlight with the window facing south. Put the other carton right next to it. Both cartons should be positioned so that their thermometers point in a northerly direction.

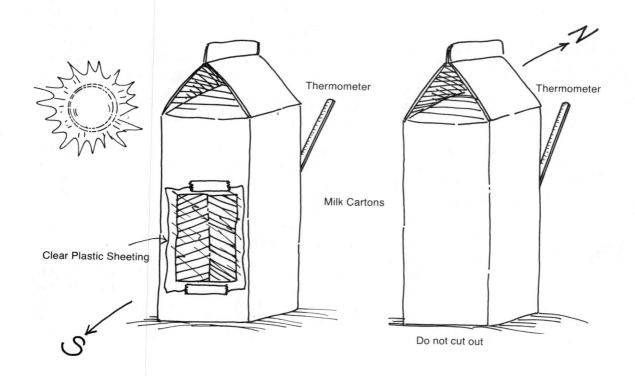

Observe and record the temperature after ten, twenty and thirty minutes. What can you infer? How is this process affected when you compare night to day, and summer to winter?

Add a roof overhang to the milk cartons and repeat this activity. Record the interior temperatures within the two redesigned (overhanging roof) milk cartons. How do these differ from your original temperature readings?

The position or orientation of a house to the sun and wind affects energy efficiency. The southern exposure should be used to capitalize on as much solar heat as possible. Future homes relying on solar energy will require planning and the reserving of an open southern exposure area. A well-designed roof overhang can protect a southern wall from summer sun, and permit the wall to be warmed in winter when the sun travels in a lower arc across the southern skies.

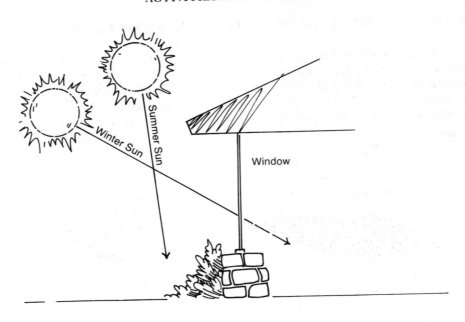

HOW DOES THE ORIENTATION OF A HOUSE AFFECT INDOOR TEMPERATURES?

Place a thermometer against a window of each side of your home or school—a northern, southern, eastern, and western exposure window. Position the thermometers in approximately the same place on each window. Record the reading for each thermometer at one hour intervals throughout the day. How do these readings change throughout the day and how do they compare to one another? What explanation can you provide for this phenomenon? Will your data vary at different seasons of the year?

Trees are helpful in both winter and summer. Deciduous trees planted on the northern and western exposures of a house save energy by shading the house during the heat of the summer. In the fall and winter seasons, the bare branches of the deciduous trees allow sunlight to reach the house and warm it.

When planning a home, build your home only as large as you need it. Heating and cooling unused space wastes energy. An energy efficient home is one with a minimum of outside surface exposed. Sprawling, one-story designs are not as efficient as a two-story house with a square floor plan. The southern and eastern portions of the house should be used for living space. Bedrooms should be placed on the northern side of the house. Rooms where heat is generated, like the kitchen and laundry room, should be located away from the western exposure.

HOW DOES SUNLIGHT AFFECT TEMPERATURE?

Pour equal amounts of cold water into two styrofoam cups. Insert a thermometer in each cup. Place one in direct sunlight. Place the other in the shade. Observe and record the temperature of each thermometer after five, ten, and fifteen minutes. Graph your results. Infer how your data would have changed if you had had different weather conditions. Do the experiment on a cloudy day, a sunny day, a warm day and a cool day and compare the results to your inferences.

CAN COLOR AFFECT TEMPERATURE?

Fold a piece of black construction paper so that it forms a small envelope. Do the same thing with a piece of aluminum foil, but make sure that the shiny side is on the outside. Place a thermometer inside each envelope and place both of them in direct sunlight. Observe and record the thermometer readings every two or three minutes. What can you infer from your observations?

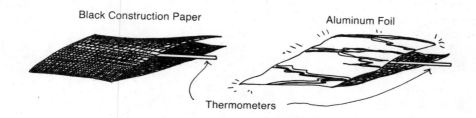

Black Construction Paper   Aluminum Foil

Thermometers

Materials Needed

3 ice cubes of equal size and shape

3 pieces of construction paper, each a different color (white, black, and one other color)

window sill or table that gets full sunlight

Procedure

Put the three pieces of construction paper side by side on the window sill or the table, in full sunlight. Simultaneously place one ice cube in the center of each piece of paper. Which melts first? Would the results be different if the paper was put on top of the ice cubes? Should the roof of a house be a certain color? What color? Should it be the same color in winter as in summer?

More Facts About Using Alternative Energy Sources in the Home

Is the self-sufficient family the answer? The rising costs of the diminishing fossil fuel supplies have caused some members of our society to alter their life style so that they are almost totally independent of standardized energy requirements. They have no requirements for natural gas or electric lines coming to their homes. They grow their own food and prepare their own dairy products. Greenhouses adjacent to their houses collect energy from the sun and store it until it is used to heat the homes. This greenhouse-stored energy provides about 50% of their heat, while a wood-burning furnace provides the rest.

A set of 12-volt batteries provides all the electricity they need. The batteries are charged about four days a month by a generator which is powered by propane gas. The TV tube is changed to run on 12 volts. A special device changes a 12-volt direct current into 115-volt alternating current to run appliances like the electric mixer, stereo, vacuum cleaner, and popcorn popper.

Is the self-sufficient family going to become the American dream? Or is the American dream going to become the desire to be a self-sufficient family?

What Will Our Energy Future Be Like?

Will We Return to Using Wind As a Viable Energy Source?

Wind is a result of the sun. As heated air rises, cooler air moves in and is directed and diverted by the spinning of the earth, topography, and the varying rates at which heat is absorbed and released from plowed ground, lakes and parking lots. For centuries humans have built mechanisms that put the power of the wind to work. Wind has moved ships, pumped water, and ground grain for us. In the decades immediately following World War I, a new type of windmill began to dot the farmlands. It was two- or three-bladed, shaped much like an airplane propeller, and positioned alongside the water pump. This windmill was used to generate electricity.

Today, larger windmills with the capacity to generate the amounts of electricity necessary for use by electric utilities are on the way. Some of these modern windmills have a span of more than 125 feet across their twin blades. The larger turbine windmills span nearly 200 feet; the approximate size of a 747 airplane's wing. These turbine windmills will be mounted on 150-foot towers and are capable of producing 1.5 megawatts.

Wind differs from most other alternative energy technologies because it is an old and proven source of energy. Part of the reason we stopped using it was our hopeful reaction to the overly optimistic projections of cheap nuclear power and the unrealistic energy prices of the 1950s and 1960s.

The basic configuration of windmills has remained unchanged for centuries. All windmills have a number of blades designed to catch the wind and thus turn the horizontal shaft they are attached to. This turning provides mechanical energy to pump water, or turn a generator that produces electricity.

The power output of a wind machine depends both on wind velocity (or speed) and on the diameter of the blades. For maximum power, a wind machine should have the longest blades possible (about 100 feet is the maximum now) and should be located in the strongest winds. The height of the wind machine is another important factor because there is greater wind speed and constancy at higher altitudes than at the earth's surface. The expense of building a tower increases with height. Most present designs call for a tower between 100 and 150 feet high.

DID YOU KNOW?

Did you know that about 2% of all the solar radiation that falls on the earth is converted to wind energy in the atmosphere?

DOES IT MAKE SENSE?

Does it make sense to allow wind energy greater than fourteen times our energy demand to go unused?

DID YOU KNOW?

Wind energy could provide 5 to 10% of our total energy needs by the twenty-first century.

DID YOU KNOW?

Winds within 80 meters of ground level possess five times the total energy we now use.

Due to the variable nature of wind energy, the major deterrent to wind power is the relatively high cost per unit of power output from the wind generating machines. On the average, "energy winds" — those with speeds of 15 to 25 mph — blow only about two days out of the week in the United States. The more common winds, called prevalent winds, blow the other five days and have a speed of 5 to 15 mph. Wind power generators operate at their rated capacity only when the wind is blowing at or above some minimum speed. Thus, a typical wind power generator can only produce about one-fourth of the total energy produced by a conventional plant of the same generating capacity.

Wind variability also has an effect on rotor revolutions per minute, which in turn influences the output frequency of the generator. Some means need to be found to regulate the frequency. This adds to the expense.

What do you do when the wind stops blowing? There is a problem in storing wind energy. Storing the energy in lead-acid batteries can cost as much as the wind machine itself. Pumping water into a reservoir and letting it fall down to turn the turbines as needed (as is presently done in conventional power plants) may be one solution. Another suggestion is to use the wind energy to compress air, store it in tanks or caves, and then feed it into gas turbines as needed. Another idea is to use flywheel storage or to electrolyze water into hydrogen and oxygen. We could then store the hydrogen for use as a fuel, or convert it to electricity in a fuel cell.

Find out additional ways that man puts wind to work. Using library facilities, find out more about NASA's windmill experiments. Also, find out more about the Smith-Putnam generator on Grandpa's Knob in Central Vermont in 1945.

Some of the many sites where large amounts of wind energy are available are: the Great Plains, the eastern foothills of the Rocky Mountains, the Texas Gulf Coast, the Green and White Mountains of New England, the continental shelf of the Northeastern United States. To produce 20% of today's power

needs or 5% of the requirements in the year 2000, we would need hundreds of thousands of the largest wind turbines presently conceived. The land requirements and aesthetics of these several-hundred-foot towers may be unacceptable to many people. The best promise may lie in use by smaller utilities, industries, and homes, where wind power has the potential to make a significant contribution.

MOVING AIR

To illustrate the effects of the heating and cooling of the earth, you need a cardboard box (almost any size will do; 30 cm by 30 cm by 50 cm is a good size), clear plastic wrap, masking tape, a large air piston syringe, heavy cotton string (12 cm long), matches, scissors, baby-food jars, ice cubes, hot water, and an aluminum pan.

Remove one side of the box, and cut a window in two of the remaining long sides. Cut out two-thirds of the top, so that one-third of the top serves as a partial roof. Tape plastic wrap over the windows so that they are airtight. In one end of the box, cut a small hole just large enough to insert a plastic straw.

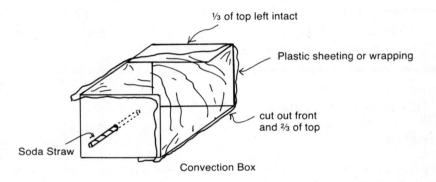

Convection Box

Cut a plastic soda straw into 4 to 5 cm lengths. Cut the heavy cotton string into lengths about 4 cm long. Double up one of the pieces of string twice or more until it fits snugly into the end of the piece of the plastic straw. Leave about ½ cm of the doubled string sticking out of the straw. Repeat this procedure for the other pieces.

Slip the other end of the soda straw onto the air piston. The air piston plunger should be depressed (inside the tube). Light the string carefully to make sure you do not melt the straw. Collect smoke in the air piston by slowly drawing out the plunger. Remove the straw and lay it aside where it won't burn anything.

Next place a pan of cold water (ice cubes in water works best) inside the convection box (by opening the plastic wrap and then taping it back down). Be sure that another straw is in place through the end of the box. The end of the straw should not be over the pan of water.

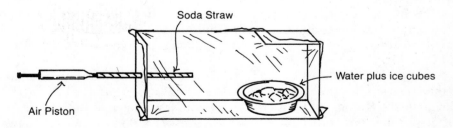

Insert the smoke-filled air piston into the straw of the convection box. Gently force smoke through the straw into the box so that it moves very slowly over the cold water. Observe what happens to the smoke as it moves into the cooler region. What do you observe? In which direction did the smoke rise? Did it fall in the direction of the pan of cold water?

Repeat this activity but substitute a pan of hot water for the cold. Again, what do you observe? Does the smoke move in the same direction? Did it rise above the pan of hot water? What is your conclusion?

This vertical (up and down) movement of air over bodies of water that have different temperatures helps to produce the motions of the wind.

SURFACE TEMPERATURES

This activity illustrates how the surface of the earth affects how air is cooled or warmed.

Materials Needed

5 styrofoam cups
5 thermometers
scissors
one 150-watt bulb with holder
water at room temperature
dry and wet sand
finely crushed dry and wet charcoal

Procedure

Cut off the tops of the five cups about 3 cm from the bottom. Save both the tops and bottoms of the cups. Fill one cup with room temperature water, one with dry sand, one with wet sand, one with dry crushed charcoal, and one with wet crushed charcoal. Arrange the cups in a circle.

Place a thermometer in each container. Each thermometer bulb should be covered by no more than ½ cm of material.

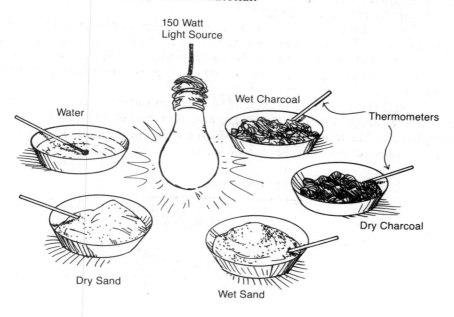

150 Watt Light Source

Water

Wet Charcoal

Thermometers

Dry Charcoal

Dry Sand

Wet Sand

Suspend a 150-watt bulb about 30 cm above the center of the circle of containers. Before turning on the light, record the initial temperature in each cup. After the light is turned on, record the temperatures after one, three, and five minutes have elapsed. Then, five minutes after the light is turned off, record the temperatures again.

FIGURE 1

Material	Thermometer Reading °C				
	Light Off At Beginning Of Experiment	Light Turned On			Light Off After Five Minute Cooling
		After One Minute	After Three Minutes	After Five Minutes	
Water					
Dry Sand					
Wet Sand					
Dry Charcoal					
Wet Charcoal					

Graph the data from Figure 1 on graph paper similar to Figure 2.

FIGURE 2   TEMPERATURE GRAPH

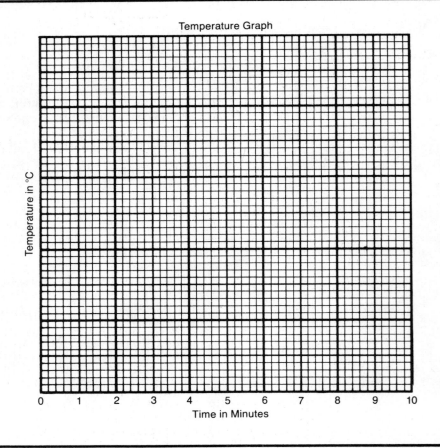

Temperature Graph

On the graph, each of the five sets of data should be represented by a different colored or different patterned line, for example, water $----$, dry sand $++++$, wet sand $0-0-0-$.

The light shone equally on all materials for the same period of time. Yet the temperature of some materials increased more than the temperature of others.

Which of the dry materials showed the greater temperature change, the dark (charcoal) or the light (sand)? Which one cooled faster? Which of the wet solids showed the greater temperature increase? Did the dry solid show more temperature increase than the same solid when wet? Did the temperature of the water increase as much as the temperature of the solid? When the light was turned off, which of the substances cooled most rapidly in the five minute period?

Rank the materials from greatest temperature change to the least. Which material changed the most? Which one changed the least?

This activity showed you that when heat from a light reaches a surface, the temperature of that surface increases. The activity also revealed that different kinds of surfaces show different rates of heating and cooling. Remember how the smoke moved differently over the cool and warm water surfaces? The different temperatures of solid surface also affect the movement of the air. This activity has just shown us how the different surfaces of the earth play a distinctive role in the development of winds.

If possible, go outside and take temperatures 2 to 3 cm above different surfaces of the earth. Are your results comparable with those in Figures 1 and 2?

MAKE YOUR OWN WIND VANE

In order to measure the wind as an energy source you need to make two devices—a wind vane for direction and a windmill for speed.

Materials Needed

1 empty spool
1 soda straw
1 straight pin
1 pencil
clay
a 4 inch cardboard triangle
thread

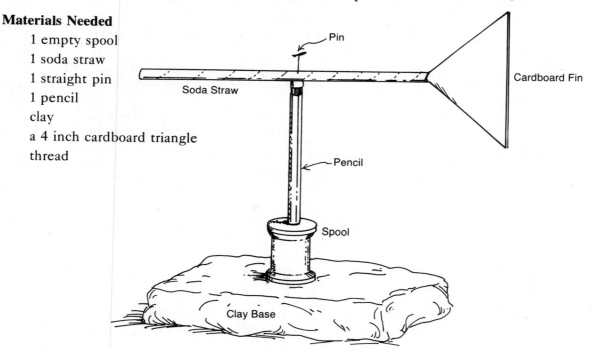

When completed, your wind vane should look like the above drawing.

Place your wind vane outdoors far away from any building. (The wind may alter direction if it hits buildings and make your data from the wind vane inaccurate.) Observe the way your wind vane suddenly moves when the wind shifts direction. To observe the variations in wind direction, prepare a chart and record the direction of the wind at the same time each day over a two week period. What is the prevalent wind direction for your area? Would you infer that it changes during different times of the year, for example, fall, spring?

WINDMILL POWER

You can make a windmill from a pinwheel, thread, paper clips, and a shoebox. Either a windy day or an electric fan can serve as your energy source. (*Caution:* an electric fan should only be used under the direct supervision of parents or the teacher.)

Color one of the pinwheel vanes a different color from the rest and measure the number of revolutions it makes per minute during a known wind speed. Once your windmill has been calibrated, you can take it outside to measure the winds over a two week period. What is the average wind speed in your area? Would you recommend your area for wind generators as an energy source? Where would you locate a windmill in your area? What would you use for energy on a calm day?

Cite several advantages of windmills. Cite several disadvantages.

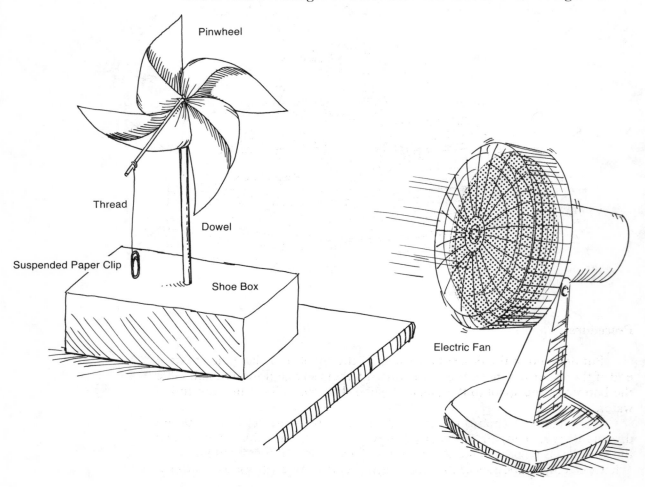

Pinwheel

Thread

Dowel

Suspended Paper Clip

Shoe Box

Electric Fan

Measure the force of the wind by suspending a paper clip from a thread as shown in the drawing of the windmill. To find the strength of the wind add paper clips until they cannot be moved by the wind. How many paper clips strong is the wind in your area? Prepare a chart to show how the pinwheel responds to various wind speeds.

Bring in a picture of a windmill. Discuss what it is and how it operates. Visit a farm to observe a working windmill if one is available in your area.

SAILING ALONG

Six thousand years ago the Egyptians put sails on their boats so that the wind could blow them from port to port. You can make your own sailboat to observe how the wind serves as a source of energy.

Materials Needed

1 piece of stiff paper
1 paint brush handle or dowel
1 spool
glue
1 block of wood pointed at one end

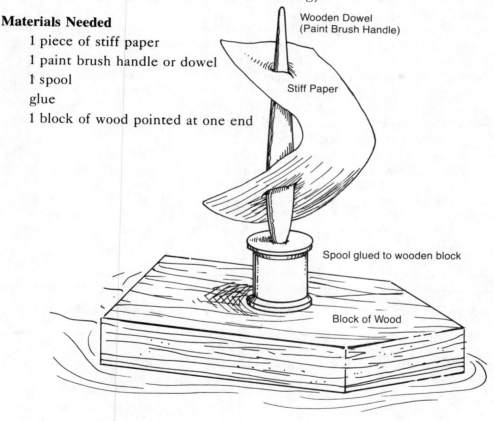

Wooden Dowel
(Paint Brush Handle)

Stiff Paper

Spool glued to wooden block

Block of Wood

Procedure

Put one end of the dowel in the hole of the spool. Punch a hole at either end of the paper and insert the dowel through them so that it forms a sail. Glue the bottom of the spool to the block of wood. Put the sailboat in a tub full of water.

What happens when you blow on the sail? Blow on the sail from different directions. What happens to the boat? How does the strength of the wind affect the speed of the boat? Make some more sailboats and conduct a sailboat race. What were the properties of the winning sailboat? Why don't we use sailboats today to move people and products across the ocean instead of just for recreation?

Find various designs of toy sailboats that can be made by children. Everyone can make his or her own sailboat and predict how it will function in the water. Test the sailboats in a large tub filled with water with a small fan as your wind source. Which designs are best and why?

Can We Extract Energy From Waste?

The concept of energy recovery from waste is not new. Paleo-indians used buffalo dung as a fuel to warm their camps and to cook their food. The low cost and seemingly inexhaustible supply of fossil fuels made the consideration of waste materials as an energy source somewhat remote. But times have changed. Waste is now rapidly becoming a valuable commodity.

Waste is generated as solids, liquids, and in gaseous forms. The latent energy inherent in waste can be recovered from the original material or from a conversion product. Most of the time energy can be obtained from waste without any alteration to its original form. The large amounts of expended gas from chemical processing industries (which use coal) can be combusted. The high temperatures that result from the combustion can be used to heat boilers. Re-using emissions is a more desirable and profitable alternative to exhausting them into the atmosphere.

Municipal solid waste combustion in on-grate, mass-burning, water-walled incinerators offers an immediate solution to waste disposal and partial energy needs of many municipalities. This energy recovery and waste disposal process has some drawbacks such as: reduced thermal efficiency due to the water content of the waste fuel, the high furnace volume to accommodate municipal waste for a given heat release, and the high cost of controlling emissions.

DID YOU KNOW?

The net yield of energy per ton of dry organic matter is equivalent to 1.3 barrels of oil.

Organic waste has the advantage of being a potential energy source which, in a sense, has already been paid for by the energy required in the growth process of the organic material.

There are millions of tons of industrial wastes, forest residues, and municipal solid wastes. The total animal manures and crop residues in the United States alone may amount to as much as eight to nine million tons of dry material annually. All of this material has great energy potential. However, these waste materials are widely scattered and the largest drawback to their use is the collecting and assembling of these materials at an energy conversion plant.

Creative thinking has spawned a good idea for manufacturing a needed product from what was once deemed useless. Industrial roof decking is being manufactured from Indiana's supply of red oak trees that once were thought to be too small and scrubby to use for anything. All parts of the formerly useless red oak tree are being used in the product. The process uses the twigs, limbs, and bark that are normally left remaining in the forest. Now, no garbage remains in the forest. All parts of the tree are chopped into small slivers and embedded in an adhesive material. The end material is red-oak particle board. It is strong and stiff and reasonably lightweight.

WASTE NOT, WANT NOT

Find out what industries are doing in your area to make needed products from "useless" materials. Have a representative from that business or industry show the class the product or arrange a plant tour.

DID YOU KNOW?

Forty percent of the electricity used on the island of Hawaii comes from the burning of sugar cane wastes. The heat from the burning creates steam which drives generators.

Stockyard animals raised in limited or restricted areas account for about half of the estimated 350 million tons of animal manure produced yearly in the United States. Useful methane gas can be produced by the anaerobic fermentation of these animal manures. One ton of animal manure can yield a volume of gas equivalent to about 4.5 gallons of gasoline. The remaining product from this process does not lose its value as a fertilizer and can be applied as soil manure.

DID YOU KNOW?

A recovery plant in Oklahoma converts the waste from one hundred thousand feedlot cattle to methane gas for commercial purposes. Decomposition releases as much as 1.6 million cubic feet a day of ammonia, carbon dioxide and methane.

The manufacture of liquid fuels from organic matter appears to be technologically possible but is not totally economically attractive yet. The gap, however, between production and cost is getting smaller and smaller. Technological developments are needed in all areas of energy development from organic waste.

Will Oil Shale Be Our Energy Salvation?

Oil shale is widely distributed. However, the deposits in Utah, Colorado, and Wyoming are of particular note because of their concentration in specific areas and their richness in organic material. These deposits are a potential source of vast quantities of liquid fuels.

DID YOU KNOW?

Estimates indicate that Utah, Colorado, and Wyoming hold as much as two trillion barrels of oil in oil shale. One third of this, or almost seven hundred billion barrels, is believed to be recoverable.

The shales of this western area were formed fifty million years ago when sediments of living organisms and precipitates of minerals collected at the bottom of large water areas. Subsequent compaction of the sediments removed the water and preserved the organic material. The high organic content of this shale indicates the prolific number of living organisms that must have characterized these waterways.

Mineralogically, the oil shales range in composition from rocks composed of illite clay to those composed of calcite and dolomite. The organic material of these shales is fairly high in hydrogen. About 65% of the shale converts to oil, 10% to gas, and 25% to a carbon-based residue. The oil yield from these western oil shales can range from 10 to 75 gallons per ton of shale.

There are two methods currently used to recover oil from oil shale. The traditional method involves the mining, crushing, and above-ground retorting of the shale. An alternate method is the "in place" or "below ground" processing technique. Because the traditional method of recovery makes up 60% of the cost of producing shale oil, the in place processing is being thoroughly investigated. In place oil shale processing involves drilling a number of predetermined wells in a specific pattern. The shale is fractured to increase permeability and then it is ignited at one or more centrally located wells. Compressed air is pumped down these ignition wells to support the combustion of the oil shale and force the hot combustion gases through the fractured rock. The solid organic material is converted to oil and the oil is then released through other wells. This method is not without problems, but it is neat and tidy. It is applicable to deposits of various thicknesses, grades of shale, and is independent of the quantity of overburden. Also, this method is highly desirable because there is no spent shale waste disposal problem.

Oil shale can be a partial solution to the energy problem. Extracting oil from shale requires highly developed technology, plus the investment of large sums of money to initiate and sustain production.

Can We Recover the Oil in Tar Sands?

We have seen that oil can be found in liquid form, and trapped in shale. It can also be found in tar-sand deposits where it binds grains of sand together. The sands must be processed in the same manner as shale — mined and then heated.

The Athabasca tar sands in Alberta, Canada have been estimated to contain approximately seven hundred billion barrels of oil. These particular tar sands cover an area of about 35,000 square miles. It is estimated that about half of the estimated seven hundred billion barrels of oil is recoverable in the form of crude oil. The oil (or tar) sands are discontinuous in nature — they vary in thickness and depth below the surface of the ground, and in oil saturation and sand particle size. All these factors bear on the economic feasibility of developing tar sands. The most important economic considerations are: the depth of the tar sands below the surface of the ground; the thickness of the tar sand layer; and the concentration of bitumen, or tar.

DID YOU KNOW?

In 1975, the Athabascan tar sands in Canada produced on the average of 42,500 barrels of oil per day.

Extracting and processing tar sands is difficult. The properties of the tar sands that cause oil extraction problems are non-uniform — varying in particle size, oil saturation, moisture content, and temperature. These variations make the development of large commercial plants very difficult.

Commercial recovery of oil from tar sands is currently being done by open-pit mining the sand and transporting it to a separation plant. This process is estimated to recover only thirty-six billion barrels of the original estimated recoverable oil (three hundred and fifty billion barrels). The vastness of the yet untapped oil continues to drive researchers on to discovering alternative techniques for extracting this oil.

In the United States, about a dozen companies are experimenting with ways to extract oil from tar sands. Tar sand deposits in Missouri, Kansas, and Oklahoma are estimated to contain more than one hundred billion barrels of oil.

Will Earth Heat (Geothermal) Be Used in Our Future?

The core of the earth is believed to be molten-rock hot. Some of this heat is pocketed in reservoirs of steam very close to the earth's surface where it escapes in the form of geysers. In other places, hot water below, but close to, the earth's surface can be tapped and changed into steam. The steam from both these sources can be used to turn conventional steam turbines to generate electricity. This system is at present being used in California, Mexico, Japan, Italy, and New Zealand.

The change in temperature from the earth's hot core to its cooler crust is also a potential source of usable energy. At about 30,000 feet (about 6 miles or 10 kilometers) below the earth's crust, the temperature differential becomes great enough to be a useful energy source. This type of geothermal energy, called "hot rock," is appealing because it is boundless. The problem is how to get at it with the least expenditure of money and existing energy, and the least impact on the environment. The depths are just now coming within present range of deep drilling technology. However, other engineering problems such as how to fracture the rock formation in the wells exists.

Of the three types of geothermal energy—dry steam, hot water, and hot rock—dry steam is the most desirable, and the rarest. Hot water sources are more common, but they are more difficult to deal with. In most cases the hot water contains dissolved salts and other minerals which are not only corrosive, but can pollute the air and contaminate surface water. Hot water sources can sometimes be used directly for space heating rather than for electrical generation.

There are still a number of problems with geothermal energy. Generating plants using the earth's steam and hot water resources have to be near the geothermal site rather than where they are needed. The installations are inherently noisy, smelly, and unsightly.

EARTH HEAT, FOR BETTER OR WORSE

Using references in your library or resource center, find additional information on and pictures of geothermal energy usage to share with your class.

Carefully boil a pan of water on a hot plate and observe how the water boils and then escapes as steam. Relate this process to dry steam and hot water geothermal energy.

Obtain pictures of geysers that illustrate what happens to geothermal energy when it reaches the earth's surface. Try shaking up a bottle of soda pop to illustrate a geyser effect.

To understand the problem of dissolved salts and other minerals, dissolve as much salt as you can in hot water. Then boil the water until it all evaporates and observe the residual materials. Relate this to the dissolved salts in hot water geothermal energy.

Obtain two pieces of pipe from a plumbing shop. Find an old, used pipe, one that shows mineral build-up in the pipe, and a new pipe. Show these pipes to your class and compare and contrast them. Apply your conclusions to the problem of dissolved minerals in the plumbing of a geothermal generating plant.

Figure 3 contains a list of locations of world geothermal power production plants. Plot these locations on a world map. Do you notice any trends? Are the majority of these locations on earthquake and volcano zones? Where do you think additional searches for geothermal energy should be conducted?

FIGURE 3   WORLD GEOTHERMAL POWER PRODUCTION*

Country	Field
United States	The Geysers, California
	Imperial Valley, California
Italy	Larderello
	Travale
	Monte Amiata
New Zealand	Wairakei
	Kawerau
Japan	Matsukawa
	Otake
	Onuma
	Onikobe
	Hatchobaru
	Takinone
Mexico	Pathé
	Cerro Prieto
El Salvador	Ahuachapán
Iceland	Namfjall
	Krafla
Philippines	Tiwi
Soviet Union	Paughetsk
	Paratunka
Turkey	Kizildere

*Source: Muffler, L. J. P., 1976, *Summary of Section I: Present Status of Resources Development:* Proc. 2nd United Nations Symposium on the Development and Use of Geothermal Resources, San Francisco, CA, May, 1975, p. XXXIII– XLIV.

DID YOU KNOW?

Geothermal energy will contribute about 3% of our energy needs by the beginning of the twenty-first century.

DID YOU KNOW?

Fifty countries throughout the world are currently active or interested in geothermal energy exploration.

DID YOU KNOW?

The largest geothermal project in the world is located about 90 miles north of San Francisco in Sonoma County, California. This project, called The Geysers, will serve more than two million people by 1985.

Can Thermal Gradient Be Trapped?

The earth receives heat from the sun and while some is absorbed, most of it is radiated back into space. This, however, is not the earth's only source of heat. Heat conduction from the earth's interior is approximately one-millionth to two-millionths of a calorie per square centimeter per second. In comparison to the sun's incoming radiant energy, which is approximately 2 calories per square centimeter per minute at the outer limits of the atmosphere, the earth's interior heat would appear to be small. Nonetheless, the earth's heat exists as an inexhaustible source of constant heat energy.

Measurements of rock temperature in mines and bore holes show that the rate of temperature rise, or thermal gradient, averages about 1 degree Fahrenheit per fifty feet (1 degree Celsius per 30 meters) of depth. This rate of increase is not constant to the center of the earth. The rate of temperature increase noted near the surface of the earth falls off rapidly with depth. It has been suggested that some of the near-surface heat is being given off by radioactive minerals that exist in the shallow layers but which may not be plentiful farther down.

FIGURE 4   TEMPERATURES WITHIN THE EARTH

Depth in kilometers	Temperatures	
	Degrees in Fahrenheit	Degrees in Celsius
30	900	500
100	2000	1100
200	2600	1400
1000	3200	1700

Although the earth's internal heat is brought to the surface in several ways—volcanic activity and heat loss through the energy used in the elevating of continental land masses—direct conduction upward through the rock is the principal mode of the heat's escape.

There is heat energy at the earth's interior. So far, we have made little use of thermal gradient as a form of energy. Suitable uses for it need to be found.

Will the Tides Aid Us in Our Energy Future?

Energy can be developed from the flow of ocean tides. Moving water possesses a great deal of potential energy, and for many years man has used the energy of waterfalls to provide electrical energy. Dams have also been used to obtain energy from artificially constructed waterfalls. Since tides are moving water, obtaining energy from tidal flow is essentially equivalent to obtaining it from waterfalls, with the exception that the water power from falls originates in a one direction stream flow.

Tides are wave motions that occur regularly, although with varying characteristics from time to time and place to place. Tidal flow consists of the vertical motion of the rise and fall of the water level, and the horizontal movement in the tidal currents toward and away from shore. Such movements are regular, and have a frequency of about twelve and a half hours between successive high tides. The movement of the tides depends on many factors, the most important of which is the changing positions of the moon and the sun with respect to the earth. Another determinant of tidal movement is the tide's geographical location.

A tidal current that moves toward the shore is called a flood tide. The greatest velocity of the moving water occurs at the approximate midpoint of the flood. The flood tide then eases off, a high tide point is reached, and then the tide begins to ebb as the water flows out. Finally, a low-tide point is reached, and the cycle begins again. The vertical distance between the high and low tide marks is called the tidal range. In some places in the world, the tidal range is as high as 50 feet; however, on open seacoasts and islands they are usually as low as 2 or 3 feet. Tidal ranges are highest in areas where either the topography of the sea floor or the coastline configuration of the surrounding land areas creates a pronounced inertia effect. This usually occurs in bodies of water that are funnel-shaped or contained at one end, such as sounds, estuaries, and bays. It is in such areas that the greatest tidal ranges and strongest tidal currents are found. The pronounced floods and ebbs of these restricted areas are classified as the reversing type of tidal current. The movement of water in bays and estuaries is complex and differs from the movements of tides in the open oceans.

The tidal currents found in restricted basin environments can sometimes be harnessed to provide energy. To do this, dams are built to enclose a basin so that the water level in the basin is different than the water level in the open ocean. Tidal movement causes the basins to fill and empty. The flow of the water created by this movement is used to drive turbines which generate electrical power. The tidal range must be at least 30 feet to be effective.

The amount of energy we can derive from the tides throughout the world is insignificant compared with other sources of energy. The world's potential tidal power amounts to less than one percent of its potential water power. There are a limited number of sites available for development of tidal power, and the installation of generating facilities is expensive. Tidal-energy projects may be feasible, and even useful, on a local basis, especially if industry can be built in conjunction with them. The first big tidal power plant was built at the estuary of La Rance, a small river on the coast of Brittany in France.

The Bay of Fundy in Nova Scotia, Canada, has long been considered as a site for the generation of electricity by tidal power. Studies of the Bay of Fundy began in 1944 and continued through the 1960s. Up until 1975, developing power from the tides was more expensive than the cost of using coal and oil. In 1976, three potential tidal power sites were identified: a 920 megawatt plant at Shepody Bay costing 1.75 billion dollars, a Cumberland Basin 725 megawatt facility costing 1.2 billion dollars and a Minas Basin Station supplying 3200 megawatts for 3.6 billion dollars.

Besides the large size of a tidal installation, a tidal plant has another disadvantage; the amount of power will vary with the tides. During a spring tide in the Bay of Fundy, when the water level becomes very high, vast amounts of power will be produced; during low tides less power is available. The generation period is also limited to between four to seven hours per tidal cycle. It may be difficult to match actual power production with the amount of electricity demanded at any one time; thus, some form of storage may have to accompany the generating unit.

HIGH AND LOW TIDES

Find out more about the Bay of Fundy in Nova Scotia and why it has been selected to demonstrate the use of tidal energy.

The United States Coast and Geodetic Survey has forty tide-collecting stations in the United States. Portland, Oregon; Portland, Maine; San Diego, California; Galveston, Texas; and St. Augustine, Florida all have tide-collecting stations. Careful measurements have shown that the sea level is not the same at all places. If the sea level at St. Augustine, Florida is taken as 0 cm high, then the sea level at Portland, Maine is 38 cm higher; at San Diego, California about 58 cm higher and at Portland, Oregon about 86 cm higher. Using this data, where would a good location for a tidal energy generation facility be? Cite the advantages and disadvantages of your choice.

Obtain a large cake pan or plastic flat plant container. Put sand (playground sand will do) at one end of the container as shown in the drawing. Fill the other end with water until it covers about three-fourths of the sand. Bury 6 to 8 pieces of gravel in the sand. Create the effect of wave action by gently pushing down on the wood with the palm of your hand and then quickly lifting your hand. Do this at regular intervals—every five or so seconds—to create gentle waves. Observe that there is no noticeable effect on the shoreline.

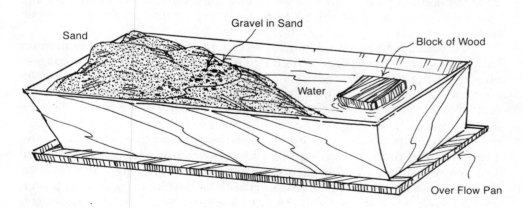

Next push down harder to create stronger waves. Observe what happens to the gravel.

Try removing some of the water to lower the sea level and try this activity again. Are there any differences in the lower sea level?

Relate this activity to the location of a tidal energy generation station and the problems that need to be taken into account with respect to the waves and changing sea level.

DID YOU KNOW?

The earliest possible data for production of electrical energy from the Bay of Fundy tides will be sometime after 1987 since a minimum period of ten years is needed for planning and construction.

How Can Ocean Thermal Energy Conversion Meet Our Future Energy Needs?

More than 70% of the solar energy reaching our planet falls on the oceans and other bodies of water. Almost a century ago a French scientist, Jacques d'Arsonval, considered the possibility of tapping the thermal wealth of the oceans. Beginning in 1926, his countryman Georges Claude—already famous for his work in liquefying gases and developing industrial uses for acetylene, helium, and neon—devoted his life to making ocean thermal energy conversion (OTEC) a reality. He succeeded in producing 22 kilowatts of electrical power at a facility built on the Cuban coast that was supplied with water from the depths through large pipes.

Claude used pumping equipment that required more power than he produced, so most engineers considered his OTEC project a failure. However, as the need for alternative sources of energy becomes more and more desirable, OTEC projects are beginning to receive additional attention.

Ocean thermal energy conversion works because solar energy heats the surface waters of oceans and large bodies of water to temperatures that are significantly higher than the temperatures of the deep waters. When the deeper, cold waters are pumped up to the warm surface waters, heat energy is released. The OTEC plant uses a working fluid, such as propane or ammonia, to be vaporized by the warm surface water. This vapor is pressurized by the heat energy and turns a turbine. The cooler water from the deep is used to condense the gas back to a liquid.

The most productive area for the OTEC idea is between the tropics of Cancer and Capricorn, where 90% of the Earth's surface is water and there is a high degree of solar radiation input. The ocean surface temperature in this region remains very close to 80 degrees Fahrenheit all year, and the underlying flow of water away from the polar ice caps keeps the deeper water at a temperature of about 35 degrees Fahrenheit. An OTEC plant would produce work from this nearly 50 degree Fahrenheit temperature difference.

There are many advantages of OTEC as an energy source. Unlike direct solar energy sources, the oceans are a huge reservoir of energy conveniently stored for use at any time. OTEC is particularly attractive for electric power generation since the oceans provide a twenty-four-hour-a-day, year-round source, rather than the intermittent supply of sunshine.

Conventional power plants produce pollution of one kind or another, and their waste products must be disposed of. OTEC is a clean source of power and can be located out of the sight of land on relatively cheap real estate.

Plants other than just power plants might be built in areas near raw materials. Aluminum plants are an example; bauxite ore could be delivered short distances to the OTEC plant by cheap water transport. There are also important by-products of OTEC plants. Fresh water is becoming as vital as energy in some areas and a 100-megawatt OTEC plant could produce millions of gallons of fresh water daily by desalting the warm water used in the plant's boilers. OTEC plants might also create productive fisheries in the waters surrounding them. Pumping cold water from the depths also brings up nutrients that attract fish to the area.

It is difficult to anticipate the problems that will be associated with OTEC plants. Some of the potential problems include: the effect of a possible change in the temperature characteristics of the tropical ocean and the resulting climatic changes. Pumping large amounts of cold water to the ocean surface would distort the ecological balance and may disrupt the natural life cycles of oxygen-producing algae. The cold water also contains large amounts of dissolved carbon dioxide which would be released upon warming. An increase in atmospheric carbon dioxide could be dangerous. OTEC plants have the potential to leak ammonia into the sea, which could endanger marine life.

By the mid-1980s the United States will decide whether or not to go ahead with prototype OTEC plants and this should determine the role of OTEC in our energy future.

DID YOU KNOW?

The Gulf Stream off the Atlantic coast of Florida has the potential to produce fifteen times the 1975 United States consumption of electric power from OTEC plants.

DID YOU KNOW?

Each year the sun bombards the earth with about eighteen thousand times as much energy as man consumes.

DOES IT MAKE SENSE?

Does it make sense to allow fifteen hundred million cubic feet per second of near tropical sea water to flow through the gulf of Florida off Florida's Atlantic coast?

SOLAR SEA POWER

Find out additional information about the French scientists who worked on the development of OTEC—Jacques d'Arsonval and Georges Claude. Also find out about the United States father and son engineering team, the Andersons, who have been promoting ocean thermal energy since the 1960s.

Locate the Tropics of Cancer and Capricorn and discuss why the area between them is an excellent location for OTEC plants.

Prove that water contains dissolved gases by filling a pan one-third full with water and heating it on a hot plate. Observe the gases moving to the surface and escaping into the air. Continue observing until all of the water is gone. Where did the water go? What are the components of water? Are they solids, liquids, or gases?

Ocean currents are caused by density differences which result in energy transfer.

Materials Needed

1 plastic shoebox
1 styrofoam cup
4 thermometers

Procedure

Fill the plastic shoebox about three-fourths full with room temperature water. Place the four thermometers at equally spaced intervals in the bottom of the shoebox. Punch holes in the bottom of the styrofoam cup with a pin. Tape the styrofoam cup in one corner of the shoebox. Read the thermometers to make sure that they all read the same. Put 100 ml of ice into the styrofoam cup and record the temperature changes shown by the thermometers. Drop small pieces of shredded paper on the surface of the water and a few soaked small pieces of shredded paper inside the cup. What happens? Try adding two or three drops of food coloring near the bottom of the cup. Did you observe currents? What caused the currents? Relate the results of this activity to the discussion about OTEC plant uses.

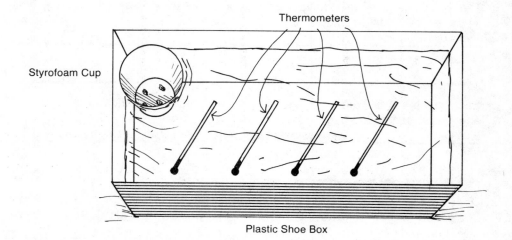

Thermometers

Styrofoam Cup

Plastic Shoe Box

Will Fuel Cells Be Our Future Power Plants?

The prototype of a new electric power option — a high efficiency, low maintenance, variable size, environmentally safe, fuel-cell generator system — is presently being developed and tested in the United States. The fuel-cell concept itself is not new — such fuel cells have already provided power to apartment houses, commercial establishments, and small industrial buildings. What is new is the effort to capitalize on the fuel cell's inherent flexibility, safety, and efficiency by using it to create a generator system that can use a variety of fuels to meet the power needs of utility organizations economically.

As indicated in Chapter 3, power generation in conventional power plants requires three steps. Only one step is required in the electrochemical process that takes place in a fuel cell. In this process, hydrogen (from coal, oil or gas) and oxygen (from air) are converted directly to electrical energy. It is this direct conversion that makes fuel cells generally more efficient than conventional generators. Fuel cells used in our space program have achieved efficiencies in excess of 75%.

A fuel cell is a sandwich consisting of an anode, electrolyte, and cathode, much like a battery. Hydrogen-rich fuel is fed down the anode side of the cell, where the hydrogen loses its electrons, leaving the anode with a negative electrical charge. Air is fed down the cathode side, where its oxygen picks up electrons, leaving the cathode with a positive charge. The excess electrons at the anode flow towards the cathode, creating electric power. Meanwhile, hydrogen ions produced at the anode (when electrons are lost) and oxygen ions from the cathode migrate together in the electrolyte. When these ions combine, they form water, which leaves the cell as steam due to the heat of the cells' electrochemical process. The fuel-cell operation is shown in the following drawing.

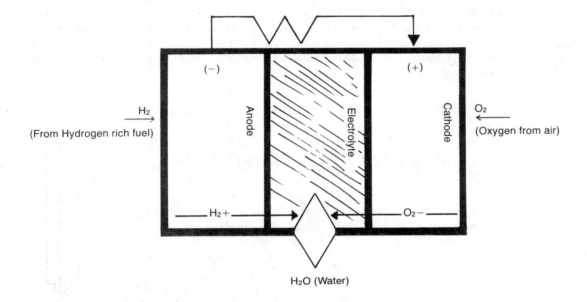

H₂O (Water)

Direct current (dc) electricity from fuel cell electrochemistry must be converted to alternating current (ac) electricity for utility applications. This conversion takes place in the power inverter which can convert large amounts of direct current power to alternating current power at nearly 96% efficiency. Emissions from a fuel cell system are ten times cleaner than that presently required by the Environmental Protection Agency (EPA). In addition to physical and operating flexibility, fuel flexibility is a major advantage of fuel cell systems. Fuel processors already accept a wide variety of hydrocarbon fuels and, by 1985, should be able to accept synthetic fuels from coal gasification.

Environmental considerations like low water requirements, limited emissions, and quiet operations help make fuel cells an attractive power option. In fact fuel cells can be air-cooled by low speed fans.

DID YOU KNOW?

Savings of one billion dollars per year (after 1985) can be expected if 20,000 megawatts of fuel-cell power are installed by 1985. This yearly savings could buy over one hundred million barrels of oil.

DID YOU KNOW?

Eighty percent of the commercial and multi-unit residential buildings annually built in the United States have a maximum power rating under 200 kilowatts. On-site fuel cells could save 25 to 30% of the fuel required to supply electricity to such buildings.

FUELING OUR ENERGY FUTURE

Have a junior-high or high-school science teacher visit your classroom with an electrolysis apparatus to demonstrate how water can be separated electrically into hydrogen and oxygen.

Using the drawing provided on p. 131, make fuel-cell sandwiches.

Find a flashlight battery and look for the cathode (+), anode (−), and electrolyte. To do this, you'll have to cut the battery in half with a hacksaw. Have your parents or teacher help you do this experiment.

Find out how fuel cells were used in our space program to send astronauts to the moon. Find out how fuel cells are used in our satellites today.

Invite a representative of your electric utility to your classroom to explain how fuel cells will be used to generate electricity for your community.

Will Hydrogen Fuel Our Future?

The perfect fuel of the future would have to be plentiful, clean, high in energy content, adaptable to power generation and to industrial, residential, and transportation uses. This is a description of hydrogen, the lightest and one of the most abundant chemical elements. Hydrogen is found in water and in all the earth's organic matter. Pure hydrogen is a clean fuel. Its only combustion product when burned with oxygen is water. It yields almost no pollutants even when burned in air.

When hydrogen is cooled to a liquid, it takes up one seven-hundredth as much space as it does as a gas. This makes it perfect for space propulsion, which requires high-energy, low-weight fuel. The space shuttle and the rockets that propelled the Apollo missions to the moon burned liquid hydrogen. Hydrogen's high energy content could also make it a desirable fuel for more ordinary transportation, and for home and industrial use. Hydrogen could help shift us away from dependence on scarcer fossil fuels.

Counterbalancing hydrogen's desirable properties is one major drawback. It is extremely rare in its elemental form. Hydrogen is almost invariably locked (bonded) into chemical compounds. Releasing the hydrogen stored in water and in organic matter requires expending significant amounts of energy.

Since energy must be invested in hydrogen before energy can be removed from it, hydrogen is considered to be a means for "storing" energy. That is, the heat or electrical energy required to separate hydrogen from the elements to which it is bonded is, in effect, stored in hydrogen until that fuel is burned.

Hydrogen was discovered in 1766 and served as a buoyant gas for balloons and as an agent for extracting metals from raw materials. In the late nineteenth century, people began to burn "town gas" or "manufactured gas": a half-hydrogen, half-carbon-monoxide fuel made from coal. Networks for distributing town gas are still in use in several countries, including Brazil and Germany.

Several million tons of hydrogen are produced annually in the United States, primarily for use in petroleum refining and in making ammonia and methyl alcohol, two major industrial chemicals. Most hydrogen is produced by reacting natural gas or light oil with steam at a high temperature. Small amounts of very pure hydrogen are produced by the process of electrolysis.

Hydrogen may now be used as a medium in which to store energy. Energy storage is vital to using energy resources wisely. Storage can make generating electricity both more efficient and more economical. It can also help us to make the best use of variable solar energy. Hydrogen could be used for both of these storage applications.

The demand for electricity produced at power plants is variable; it is higher during the day than at night; higher during weekdays (when businesses are in operation) than on weekends; and, in many parts of the country, higher during summer (when cooling needs are up) than the rest of the year. Peak demands that threaten to exceed a power plant's generating capacity alternate with demands so low that much of that generating capacity is idle. The ability to store the excess energy that could be produced during low demand times to supplement the electrical output during periods of peak demand is called "load levelling." Load levelling would enable power plants to operate more efficiently and economically. Electricity itself is difficult to store economically, but it can be converted into a more easily storable form, like hydrogen. A utility could produce hydrogen with excess electricity during off-peak times, store it, and reconvert it to electricity in fuel cells during peak demand times.

Practical use of solar energy also requires a way of storing energy for backup use when there is no sunlight. Hydrogen produced with electricity during high sunlight hours could be burned later to produce supplemental heat for solar heating systems or to drive electrical generating systems.

As an energy carrier and storage medium, hydrogen has several potential advantages over electricity and devices that convert and store electrical energy. Hydrogen can be transported long distances by pipelines instead of expensive overhead transmission lines that require wide right-of-ways. Hydrogen can probably be stored as cheaply underground as natural gas is.

Present automobiles, home furnaces, power plants, and industrial plants can easily be converted to burn hydrogen instead of fossil fuels.

The biggest roadblock to the use of hydrogen is cost. Less expensive, more efficient production, storage, and distribution methods are needed if we are to use hydrogen as an energy source. Right now, hydrogen is much more expensive than the fossil fuels it might replace. If new technologies are developed, and if fossil fuel prices continue to increase, then hydrogen may become a practical and affordable solution to our search for energy.

Researchers are working on making hydrogen from water by electrolytic and thermochemical methods that can use solar and nuclear energy. The electricity needed to electrolyze water for hydrogen production can come from today's fossil or nuclear power plants, or it could be produced with solar energy.

The sun's energy could also be used to produce hydrogen by a process called "photolysis." Photolysis is the process in which light decomposes materials. Electrodes made of semiconducting materials can absorb sunlight and split water at the electrode surface. A solar-powered "electrolyzer" based on this process would need almost nothing except electricity.

Thermochemical cycles—the reaction of chemicals at high temperatures—is another method that may produce hydrogen.

In addition to its potential use for generating electricity and its regular use for rocket propulsion, hydrogen has also been used experimentally as a fuel for airplanes, naval vessels, and motor vehicles. Hydrogen's big advantage in transportation is light weight; however, its biggest drawback is storage. Hydrogen requires four times the space jet fuel requires and twice the space fuel for automobiles requires.

In conclusion, hydrogen also has another big disadvantage to consider. It is extremely flammable and it burns with an invisible flame. Thus the public will have to be educated in the safe use of hydrogen. Furthermore, a flame colorant may need to be added to the hydrogen flame to make it visible. The potential of hydrogen as an energy source will probably begin to be utilized in the twenty-first century.

HYDROGEN FUEL

Conduct the electrolysis activity in the fuel cell section of this chapter to illustrate the properties of hydrogen.

Find out about the electrical demand in your area. What are the peak demand days, hours, etc.? Have everyone find out the peak hours of electrical use in their homes. Have your class formulate a plan for their homes so that electrical use will be reduced during those peak hours.

Investigate the famous blimp called the Hindenberg. What happened to it and why? What gas is used in blimps today? Find out about the Goodyear (tire not publisher) blimp. Compare it to the Hindenberg.

Prepare posters highlighting the advantages and disadvantages of using hydrogen as a fuel for the future. Incorporate safety suggestions in your posters.

DID YOU KNOW?

Hydrogen's energy content per pound is almost three times that of gasoline. Hydrogen has the highest energy content per pound of any fuel known.

Is Fusion Energy Our Salvation?

Fusion, which is the power of the sun and other stars, is another form of energy being developed for electrical power production in the future. The process can be safe, environmentally attractive, and, best of all, the supply of deuterium and tritium can be produced in a fusion reactor from lithium, an element found in granitic rocks and underground salt water.

Fusion is the process by which the sun generates its energy. The sun is a very hot gas; in fact, it is so hot that atoms in it have been ionized. This means that the negatively charged electrons have been separated from the positively charged nucleus (called an ion) of the atom.

This gas of electrons and ions is called a plasma and has some very special properties. One of these properties is that the ions frequently collide with each other. The hotter the plasma, the harder they collide.

If the plasma is hot enough, then the ions will collide with enough force to overcome their tendency to repel each other due to their positive electric charge. When this happens, these ions or nuclei combine or fuse to form new nuclei (new elements) and release energy in the process.

What is needed is a way to generate a very hot plasma, like the sun, and hold it together long enough for many fusions to take place and release a lot of energy. The major problem facing researchers is how to hold and heat this

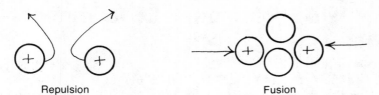

Repulsion   Fusion

plasma in a suitable container so that the ions, which are moving more than a million miles per hour, will not strike the walls and lose their energy.

One method of confinement is to keep the plasma inside a magnetic field called a "magnetic bottle." Results of experiments conducted in the mid–1970s have shown the practicality of building larger fusion machines, which can produce net fusion power. A machine that can burn deuterium and tritium will be built about 1982. The first experimental fusion power plant, which will produce electricity on a modest scale, will be built about 1987. Thus, we can anticipate the first commercial-scale fusion power plant sometime near the beginning of the twenty-first century.

First generation fusion power plants will generate electricity by means of a thermal energy conversion cycle, which will obtain heat energy from the fusion of deuterium and tritium. Tritium is radioactive and deuterium is not. In spite of the fact that tritium is one of the least hazardous of the radioactive elements, power plants that use it will have to be carefully designed to assure that the tritium will be safely contained under all operating conditions.

Eighty percent of the energy released by this reaction will be carried by neutrons. This requires that the reaction chamber be surrounded by a region in which the neutrons can slow down and release their energy as heat to a working fluid. The fluid will in turn be used to generate steam to drive an electric turbine generator.

Fusion systems have the potential to play a big role in our energy future.

FUSING ALONG

Obtain a specimen of granite. Granite contains an element called lithium that contains the fuel for a fusion reactor.

Prepare comparison charts for fission and fusion reactors and cite the advantages and disadvantages of each. Examine both reactors' advantages and disadvantages in the area of environmental effects and safety. Which type of reactor (fission or fusion) is the safest? Which type of reactor has the least problem with the storage of radioactive wastes?

Pass magnets around your class so everyone can feel the force of like charges repelling each other. Discuss the problems associated with getting like charges to go together for fusion purposes.

DID YOU KNOW?

If the energy in the deuterium in the Pacific Ocean were used to fuel fission reactors, it would provide enough energy to generate electricity for the entire world for billions of years.

DID YOU KNOW?

Studies made in recent years show that fusion systems could be used directly or indirectly for the manufacture of combustible fuels such as hydrogen, synthetic natural gas, and alcohol, from water and gases found in the air.

Will Hydropower Be Revisited?

The history of the United States reveals that the early settlers lived in balance with their environment. In the late nineteenth century, wood was the principal energy source used to heat homes. Emerging industries used a combination of wood, water, and wind as sources of energy. Water wheels were responsible for driving much of the basic machinery of New England industries. By the middle of the nineteenth century, wind and water provided the energy for 64% of the mechanical work done in this country; coal accounted for 19%; and, wood supplied 17% of the energy used in industry. These figures reversed themselves within twenty years. Coal usage increased to 58%, wind and water usage dropped to 33%, and wood usage dropped to 9%. Now, the United States is overconsuming its non-renewable fossil fuels. However, the United States could not have attained the world status it currently enjoys in the time span it did without this change of energy sources.

DID YOU KNOW?

Water power is currently responsible for about 15% of the total generating capacity of the United States.

Although water power is a minor source of today's energy, it has continued to be a consistently used source of energy. In fact, during the last few decades, the total hydro-generated electric energy has been doubling every sixteen years.

The average person uses about 200 gallons of water a day for home use. This does not include each person's share of water used for industrial and commercial uses, and public services.

MEASURING STREAMFLOW

To determine the flow of water in the stream you've chosen, measure and mark a 100 foot distance along a straight section of the stream. If necessary, a shorter distance will suffice. Cast a 4 or 5 inch long stick in the water above your upstream marker. Record the amount of time in seconds that the stick takes to float between your starting (upstream) and ending (downstream) markers. Divide the distance traveled from marker to marker by the recorded time in seconds. This will provide you with the average speed traveled in feet per second (or the flowage of the stream in feet per second). Repeat this measure several times. Determine the average speed of the streamflow.

Find the average width of your section of the stream. To do this, measure the width of the stream in a number of places between your markers. Determine the average of these readings.

Find the average depth of your designated section of the stream. Take three readings, at the same cross section and calculate the average depth. Remember the average of the three readings will give you an average, but it won't give you the average depth. To find the average depth, divide the three cross-section readings by the number 4.

To find the streamflow, multiply the average width by the average depth,

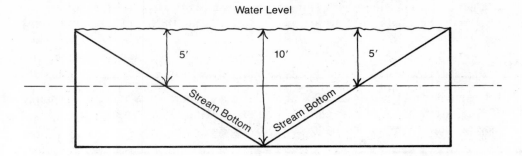

and multiply that figure by the rate of flow per second. How many people could your stream support relative to their water requirements?

If a stream is not available, use either a section of vinyl or aluminum guttering or a large, flat plastic tray for your stream bed. Use a bucket with a plastic tube for your water supply and place playground sand into the tray.

Connect a plastic hose to the supply bucket and also connect a plastic hose to one end of the stream bed. Check for leaks. Elevate the supply bucket and run the hose into the stream bed. Run the hose from the stream bed into the catch bucket. Varying the rate of water flowing from the supply will result in a variety of stream patterns. What stream bed patterns reflect a slow flowing stream? A fast moving stream?

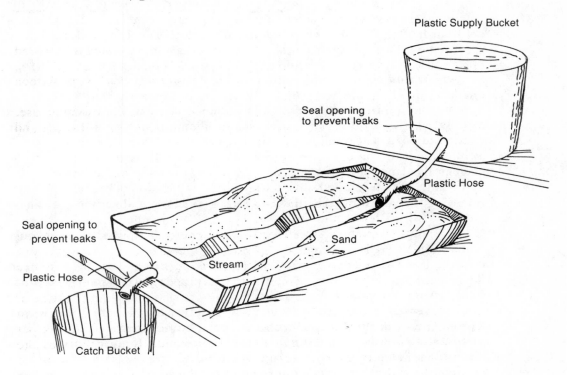

Are We Using Solar Energy to Its Maximum?

Solar energy is the energy received by the earth from the sun. The sun has provided, directly or indirectly, almost all the sources of energy for the earth since its beginning. Solar energy provides direct solar radiation for heat and for the generation of electricity. Indirectly, solar energy is responsible for hydro-electric power, wind, fuel from organic waste, ocean thermal movements, and ocean waves.

DID YOU KNOW?

It is estimated that by the year 2000 solar electricity generation will be approximately 10% of the total United States electric generation. At the same time solar energy may also account for 5% of the nation's energy needs.

Solar energy is kinetic energy. It is radiant energy, much of it in the form of visible light. The sun radiates in all directions and only about one two-billionths of this reaches earth. In three days this tiny fraction of the sun's energy provides about as much heat and light as is available from all of our known reserves of coal, oil, and gas. This tiny amount is plenty for us. If we could convert at 20% efficiency, the amount of solar energy that reaches an area the size of New Jersey could fill our current needs for electrical power.

Direct solar energy is free, non-polluting, and safe. It also is non-depletable. On the negative side, solar energy is weak, highly scattered, and extremely variable in its consistency. Utilization of solar energy requires a large collection area and means for concentrating and storing it. These processing problems presently make a free source of energy expensive. And yet, the amount of solar energy that falls on the earth globally in one day is equivalent to one-fourth of the fossil energy contained in the earth's total coal, oil and natural gas reserves. This makes this form of energy too valuable to not take advantage of.

Plants capture sunlight by the process of photosynthesis. Photosynthesis enables sunlight to provide the energy needed to convert atmospheric carbon into organic forms which other living forms use as food. Photosynthesis is a process which converts radiant energy to chemical potential energy. The energy stored in the carbohydrate can be released by burning it (or oxidizing it). Sunlight and the process of photosynthesis is a fundamental element in the ecological balance of nature. Dead and decaying plants, through the process of being inundated by eroded soils, plus the later application of heat and pressure following the burial process, turn into coal, petroleum, and natural gas. These are the fossil fuels that currently provide 93% of the energy used by man.

The most common method of using solar energy has been the greenhouse application. This is principally a glass structure that provides a controlled climate. It is an effective way to trap and convert the sun's radiation to heat. The greenhouse effect is based on the specific property of common glass which transmits the shorter wavelength visible portions of sunlight and at the same time prevents the passage of longer wavelengths, the invisible heat waves. Visible energy of the sunlight enters the greenhouse where it is absorbed by plants, soil, and the general fixtures of the interior of the greenhouse. This absorption changes the incoming sunlight to longer wavelength heat waves that are retained inside the greenhouse and raise the interior temperature. This same basic process has been applied in simple solar-heat-to-collector units.

BUILDING A SOLAR GREENHOUSE

How can knowledge of which material absorbs heat most readily and which retains it the longest (air, soil, or water) assist you in building a solar greenhouse that will maintain a steady temperature throughout the winter?

Select two cardboard boxes of equal size. The dimensions should be in the 50 cm cubed range. The overall dimensions are not critical. It is more important that you have two boxes of equal size. Construct one greenhouse in a traditional style, the other in a solar greenhouse style. The traditional greenhouse simply has the front cut out, while the solar greenhouse has the front, sides, and top cut out (see diagram). Cut the angle indicated based upon your geographical location (your latitude plus 15 degrees), cover the front of the traditional greenhouse and the front, sides, and top of the solar greenhouse with three or four mil clear plastic.

Suspend a thermometer in each greenhouse. Turn the fronts of both greenhouses directly into the sun. Take readings and record the temperatures in each greenhouse throughout the day.

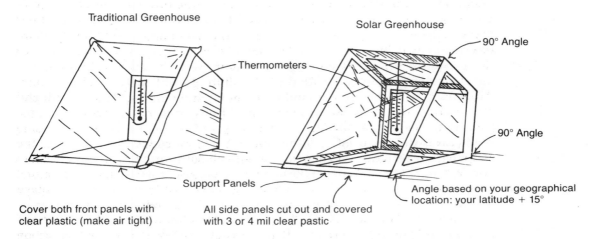

Collect an array of similar sized, used cans. Spray them with flat, black paint. Fill them with water almost to the top. Seal the openings of each can with plastic wrap held snugly around the top by a tight rubber band. Stack these cans across the back section of the solar greenhouse.

Again, suspend a thermometer in each greenhouse and place them in full sunlight. Measure and record the temperatures in each greenhouse throughout the day. Continue recording the two readings long after the sun has gone down. How do the two temperatures compare? What explanation can you give to support your comparison?

What can you do to further enhance the retention of absorbed heat in the solar greenhouse? What, where, and how would you insulate the solar greenhouse? How could paint and the color of the paint assist you in retaining heat?

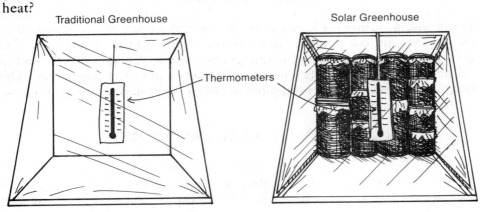

INVESTIGATE A PASSIVE SOLAR HOUSE

Materials Needed

1 cardboard box
scissors
plastic wrap
light source
white paper
black paper
thermometer

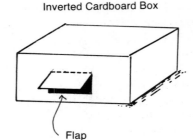

Inverted Cardboard Box

Flap

Procedure

Cut the lid off of the cardboard box. Then cut a flap in one side of the box and cover the opening with plastic wrap.

Cover the top of the box with white paper and shine a light into the box and record the temperature with the flap open and with the flap closed. Repeat this activity with the light off and record the temperatures. Then repeat the entire activity with the box covered with black paper.

- What is the best roof color for houses in cold regions?
- What is the best roof color for houses in warm regions?
- With the light on, is the highest temperature reached with the flap open or closed?
- With the light off, is the highest temperature retained with the flap open or closed?
- In the summer, should the flap be open or closed during the daylight hours?
- Should the flap be open or closed during the daylight hours in the winter?
- During the nighttime hours should the flap be open or closed in the summer?
- Should the flap be open or closed during the nighttime hours in the winter?

Relate the cardboard box to your home or school. Pretend the plastic wrap is your window area and the flap is a window shade or drape. Design a plan to keep your home or school the warmest it can be in the winter and coolest it can be in the summer.

BUILDING A PASSIVE SOLAR HOME

If you sit in front of a window on a sunny day you can feel the warmth through the window. Windows in a house are very good solar collectors. They allow the sun's energy in and trap it so that very little escapes.

In this activity you'll learn ways to landscape a lot and orient a house to achieve maximum benefit from the sun's energy without using complex solar collecting systems.

Materials needed

scissors	ruler
tape	duplicates of Figures 5 and 6.
paper	light source (filmstrip projector, lamp, flashlight)

Procedure

Cut and fold the model homes in Figure 5 and assemble them without tape. Place the folded model homes on a sheet of paper and decide on the location of each home. Make decisions about the room, windows, and door placements. Unfold each home and draw in the windows and doors. Check to make sure you want them where you drew them, and if so, cut them out.

Refold and tape each home. Then tape each roof in place and place each model home on the plot plan. Cut out the model trees and shrubs in Figure 6 and fold the bases. Tape the models wherever you think they would be best on the lot. Use as many tree models as you desire to landscape each home. Remember that deciduous trees lose their leaves in the fall.

Get the light source and set it up so that it affects your paper lot and house the way the sun hits your real house. Remember that the angles and intensity of

FIGURE 5

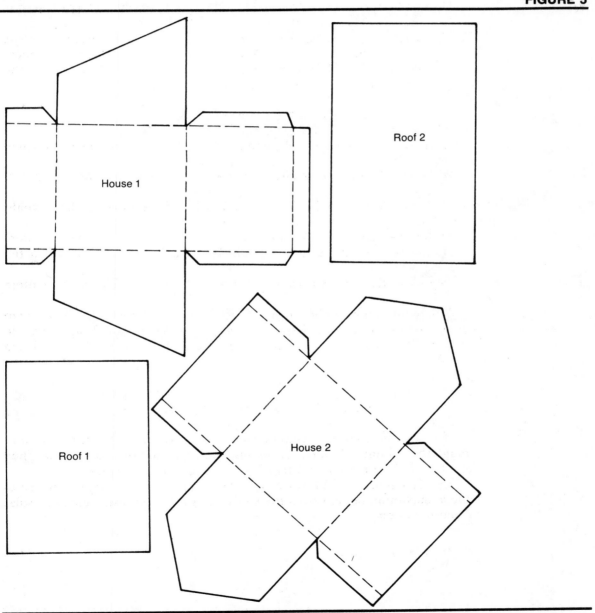

FIGURE 6

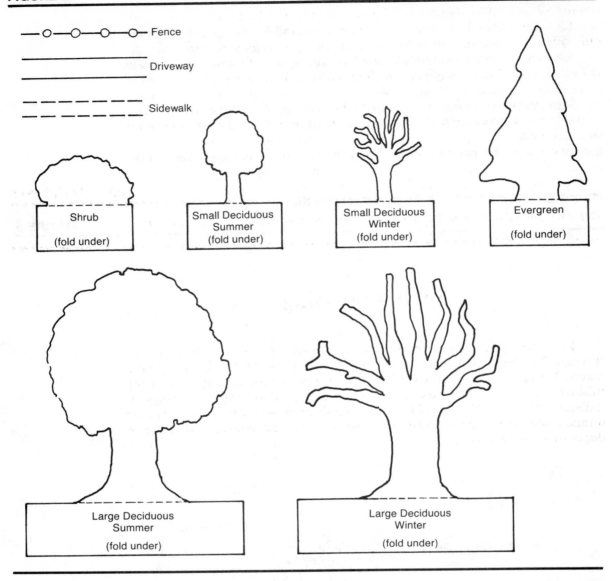

——○—○—○—○— Fence

Driveway

Sidewalk

Shrub
(fold under)

Small Deciduous
Summer
(fold under)

Small Deciduous
Winter
(fold under)

Evergreen
(fold under)

Large Deciduous
Summer
(fold under)

Large Deciduous
Winter
(fold under)

the light change in winter and summer. You'll have to set the light source up twice—once for summer and once for winter. Are your trees and shrubs providing effective winter heat and summer shade? Be sure to interchange the summer deciduous tree models with winter models.

- How does your dwelling compare to those of other students in the placement of windows, size of doors and roof arrangements?
- What are some common procedures for landscaping passive solar homes?
- In which direction should the largest roof overhang face to take advantage of winter sun while avoiding the summer sun?

Redesign and relandscape each home so that they have active solar energy heating systems. Place model flat plate solar collectors on the roof (see page 144). Determine the most efficient roof pitch for such a system and re-draw the sidewalls to obtain this pitch.

Redesign each home to include a cooling system, such as an evaporative cooling system. What factors will have to be considered for such a system?

Construct a model home with various passive solar design features in it, only this time use other materials such as wood, or plaster of paris.

Design an evaluation system to decide which home design and landscape is best. Points should be given for the following features.

- Most windows face south with a roof overhang for summer use.
- Deciduous trees are planted or oriented to the south and west sides of the house.
- Evergreens protect the house from the prevailing winds in a given locality.

DID YOU KNOW?

The first solar collector was developed and used more than one hundred years ago.

Solar Collecting

In a solar collector like the one in the drawing, the sun's energy passes through the glass cover plate. It is then absorbed by the black background material. Because the energy can't escape, the water circulating through the tubes is heated. The heated circulating water is then stored in tanks or radiators. This type of solar collector can supply buildings with hot water in the temperature range of 38 to 93 degrees Celsius (100 to 200 degrees Fahrenheit), depending on conditions.

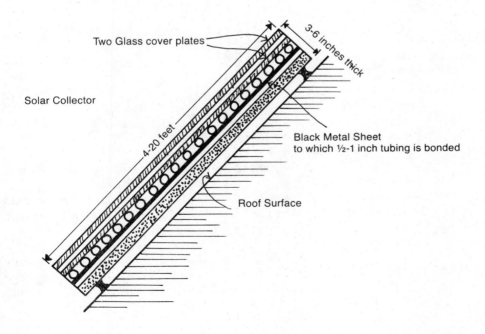

Two Glass cover plates

3-6 inches thick

Solar Collector

4-20 feet

Black Metal Sheet
to which ½-1 inch tubing is bonded

Roof Surface

Building a Flat-plate Collector

Select a piece of soft pine wood 12 inches by 5 inches by ¾ inch thick (or 30 cm by 15 cm by 2 cm). The dimensions are not critical, almost any scrap piece of lumber will do. Completely cover this block of wood with heavy duty aluminum foil. You will need 5 to 10 feet of flexible plastic tubing with a ⅜ to ½ inch diameter. Nail large fencing staples in two straight lines on the board. The rows of staples should be parallel to one another and far enough apart from each other to accommodate the interlacing of the flexible tubing. (See the drawing.)

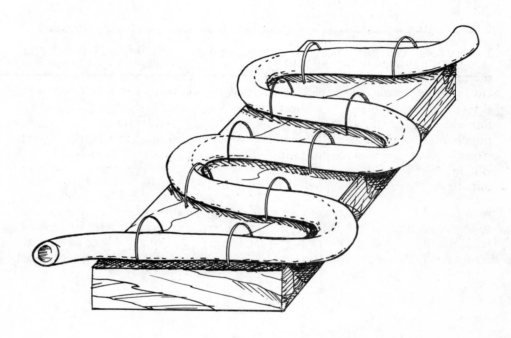

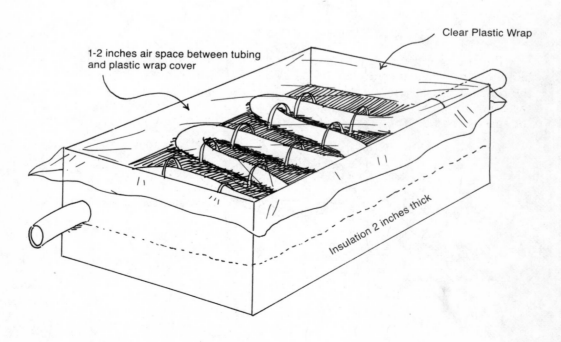

1-2 inches air space between tubing and plastic wrap cover

Clear Plastic Wrap

Insulation 2 inches thick

Lace the tubing back and forth through the staples. Make sure that the staples are not nailed too close to the board. The tubing should easily fit through the staples so that there are no kinks or folds in the tubing. Spray the entire construction with black paint. Find or construct a shallow box big enough to hold the plate collector, the 2 inch thick insulation, and 2 inches of free space. Lay the insulation in the bottom of the box. It should cover the entire bottom of the box and be 2 inches (5 to 7 cm) thick. Put the plate collector on top of the insulation. There should be 1 to 2 inches (2 to 5 cm) of space above the plate. Cover the top of the box with clear plastic wrap. Secure this snugly with tape. No air should be allowed to escape from within the system you have just created.

USING THE SOLAR COLLECTOR TO HEAT WATER

Get a 1 or 2 pound coffee can with a lid. Punch one hole about 4 cm from the top of the can. On the opposite side of the can you've just cut, cut a hole 4 cm from the bottom of the can, so you have a high and a low hole on opposite sides of the can. Insert two short lengths (approximately 3 inches each) of flexible plastic tubing into the can through each hole. Make sure that the punched holes are a bit smaller than the outside diameter of the flexible tubing. After you've squeezed the tubes into the holes, apply a water soluble glue or a silicone tub caulking compound around both sides of the holes (inside and outside of the can). This should make your can airtight.

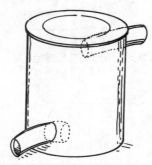

Inlet: 4cm from top of can

Outlet: 4cm from bottom of can

Put the coffee can with tubes in a large container, like a large ice cream barrel or a rectangular cardboard box. Punch two holes in the container so that the tubes in the coffee can will poke out of the container. (You'll have to pull the tubes through.) Surround the coffee can with enough insulation to fill all the space between the coffee can and its container. This is your hot water tank.

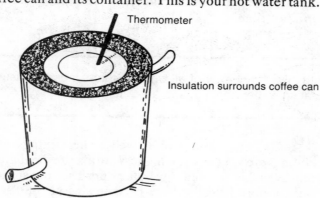

Thermometer

Insulation surrounds coffee can

Connect the tubes of your hot water tank to the extended hoses leading in and out of the collector. You can join the loose ends by inserting a plastic sleeve (plastic soda straw section) into the hoses and clamp the hoses by twisting a lightweight wire or pipe cleaner around the hose where the straw is inserted.

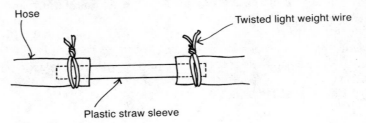

Punch a small hole in the plastic cover of the coffee can to allow a thermometer to fit down into the water. Make sure enough of the thermometer sticks out of the top of the can so that you can observe changes in temperature. After the solar collector and hot water tank have been joined together as a closed system, place the solar collector in direct sunlight with the water tank (coffee can) positioned at a level higher than the solar collector.

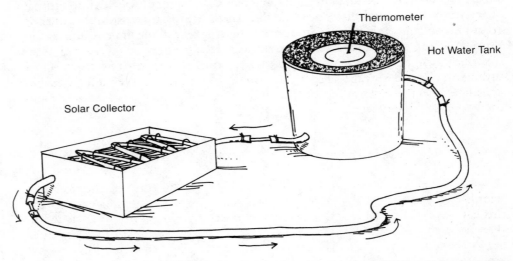

What basic principles make the system work properly? Will this solar water heater work if the water tank is on the same level as the collector? Will it work if it's below the collector? When the sun has set in the west and night falls, will the process operate the same? Will it stop? Will it reverse itself? Can you calculate the number of calories of heat produced in the system in a given period of time?

DID YOU KNOW?

Roughly one quarter of all United States energy consumption is related to space heating, water heating, and air conditioning.

Solar energy can cool as well as heat things. Cooling requirements for buildings are apt to be the highest during the daylight hours and in the summer when available solar energy is at its peak. In the summer there is not as great a need to store heat as there is in the winter. Solar cooling uses absorption refrigeration equipment similar to the type used in gas-burning refrigerators and air-conditioners used in campers. The solar heat energy simply substitutes for the gas flame to heat the refrigerant which then expands through a turbine to drive a mechanical air conditioner.

How Can Solar Energy Be Converted to Electricity?

Wind power was once used for pumping water, operating machinery, and later for generating electric power. Cheap, available electricity for rural farms was the main reason for the decline of windmills. The rising cost of electricity, however, has caused renewed interest in windmills. A wide range of studies and experiments in wind energy conversion technology is under way. We still do not know how much of the total available wind power can be fully utilized.

Through the use of lenses, curved mirrors and other collectors, we can concentrate and focus solar radiation to create heat energy. Solar collectors can concentrate enough heat to produce steam for a turbogenerator capable of generating electricity. All solar thermal conversion systems have similar basic elements. They are: a concentrator which focuses the sun's rays; a receiver to absorb this energy; a means to move this heat to a storage area or directly to a turbogenerator; a storage area to accumulate a heat reserve; and a turbogenerator which converts the heat energy of steam to mechanical energy and then to electrical energy.

COOKING WITH A FRESNEL LENS

A Fresnel lens is an optical paraboloid reflector. You can buy one from any scientific supply house. A Fresnel lens of at least 25 cm across will work best for this activity. The focal length of your lens should be listed on the instruction sheet that comes with it. If this is not furnished, place a sheet of white paper below the lens, move either the lens or the paper until you observe the smallest pinpoint of light coming through the lens onto the paper. In bright sunlight this paper should smolder, ignite, and burn. Be careful when you are establishing the focal point — be prepared to observe all normal precautions when dealing with an open flame. This establishment of a focal point should be done with a teacher or parent, and should take place outdoors in a spot that is protected from the wind. When the paper is ignited, the measured distance from the Fresnel lens to the paper is the focal point.

Materials Needed

To construct a Fresnel solar furnace or cooker, you will need: a Fresnel lens; a holder or frame for the lens; a stand; some nails; glue; and/or wing nuts and bolts; and a pair of extension arms fastened to the Fresnel lens frame so that the small metal container you are going to make can be positioned at the focal point of the Fresnel lens. All the dimensions of the solar furnace depend on the size, shape, and focal length of your Fresnel lens. Although a drawing is provided here, you will need to establish the correct and appropriate dimensions for your constructed solar furnace.

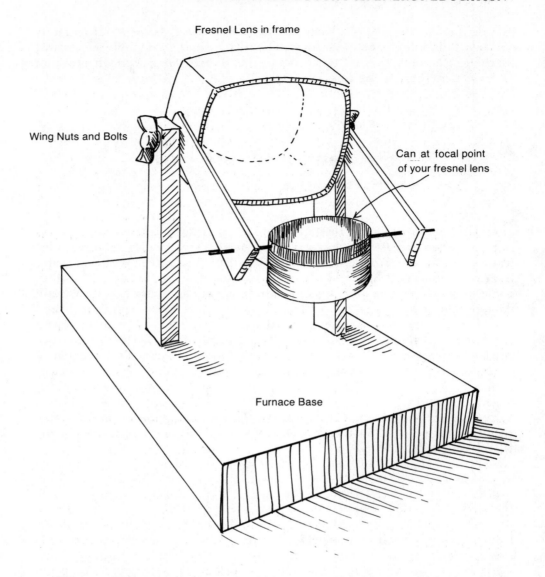

Fresnel Lens in frame

Wing Nuts and Bolts

Can at focal point
of your fresnel lens

Furnace Base

The small metal container should be approximately 1 to 1½ inches high and 2 to 2½ inches in diameter. A small, metallic, cut-down, orange juice can will do adequately. A teacher or parent should supervise the cutting of the metal juice can. When you are finished cutting, file the edges of the can smoothly. Next, drill or punch small holes in two opposite sides of the can and push a small stiff wire through the holes to create a handle (a portion of a coathanger works equally well). Paint the entire can and handle black. Now, weigh the empty can. Fill the can with water, but make sure the water level is below the punched holes. Add black ink to darken the water and weigh the can with the water in it.

Place the can with the water within the focal point of the Fresnel lens. Place the entire apparatus in sunlight. Position the Fresnel lens so that maximum sunlight is directed at the black can and its contents. Record the temperature of the water before exposing it to the sun's energy. Record it after two minutes and compute the calories of heat attained. How long does it take to boil the water using the sun?

A very promising space age accomplishment is the development of technology for converting the sun's radiation directly into electricity through the use of solar cells. A solar cell contains two very thin layers of silicon with an outside wire attached. In one layer, a few atoms in the silicon crystal have been replaced by boron atoms; in the other, the replacement atoms are phosphorus. Sunlight falling on the cell forces electrons to move along the wire from the phosphorus-silicon layer to the boron-silicon layer. Converting radiant energy into electrical energy is accomplished by striking solar radiation on certain materials, such as, silicon, cadmium sulfide, cadmium telluride, or gallium arsenide. These materials absorb the solar radiation and cause a separation of the electrons from the atoms. The movement of the electrons in one direction and the positively charged ions in the other direction produces a small potential difference, approximately 0.5 volts. An assemblage of these solar cells can produce useful electric power. This process is called photovoltaic conversion and is accomplished without the use of any other outside power source. The size and type of solar cell or solar-cell panel depends on the power needed, available sunlight, and the geographical location of the cells.

As with all new innovations, the initial cost for large central installations of solar cells was exceedingly high. Research and new advances promise to lower the cost of solar collectors.

DID YOU KNOW?

A 6 by 9 meter solar-cell panel operating at 10% efficiency (at noon in the northeastern United States), with a peak output capacity of approximately 6 horsepower (5 kilowatts), would yield an average of approximately 1 kilowatt over an entire year.

The most attractive energy solution would appear to lie in the expansion of cheap, clean, renewable solar power. However, despite recent advances, existing solar technologies can provide only limited help over the next ten or twenty years. The major problem appears to be effective storage of the heat that solar energy can provide. We must remember that all of this can change rapidly through major breakthroughs in technology. However, at the present time, it seems that central electrical power generation by direct solar energy may be unlikely. Small, self-contained energy systems combining solar generation with wind energy or energy from wastes, along with some electrical storage capacity, will probably play an extremely important role in supplying energy to rural areas or to apartments and buildings.

SOLAR DEHYDRATION

Many fruits can be dried by using energy from the sun to evaporate the water contained within the fruit. You must take several precautions if you want success in drying fruit. First, you need to construct a screened, drying tray in which the fruit can be placed, and through which air can circulate around the fruit or fruit parts. The screened drying tray must be placed in direct sunlight with good air circulating through the tray. Also, once the drying process starts, the drying tray must be protected from excess external moisture and rain. Thus, it will be necessary to cover or bring the drying tray indoors at night and during rain periods. Finally, you must also protect these tempting fruits from birds and insects.

Two small wooden frames, or two constructed from heavy cardboard, covered with plastic window screen (these are available at local hardware stores), tape and wire are all you need to make one or more drying trays. Place the window screen between the two frames and tape the edges together. Tie a

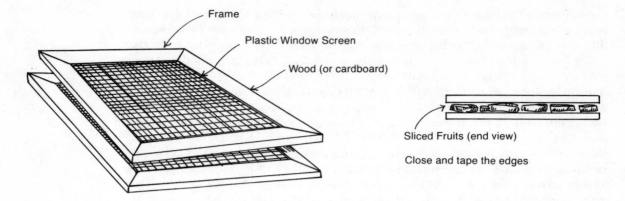

Frame

Plastic Window Screen

Wood (or cardboard)

Sliced Fruits (end view)

Close and tape the edges

piece of string around each corner (through the screen and around the frame) and then tie the ends together to make a hanger.

You can dry sliced peaches, apricots, prunes, or grapes on this screen. If you use a solid tray instead of a drying screen, you can make tomato paste from fresh tomatoes. Find out if this same process can be applied to drying fish or meat.

Hang in sunlight

Solar Energy Can Dry Crops Effectively

Corn can be harvested as soon as it matures, or it can remain in the field after it matures for several weeks before it is harvested. If corn remains too long in the field after maturation it is susceptible to mold, fungi, bleaching, general weather damage, birds, and rodents.

Modern automatic corn-pickers can operate more efficiently if the corn has not been allowed to dry in the field. However, even though crop losses are significantly reduced by this early picking, the moisture content at this time is high and must be lowered from 30% to 13% before it can be safely stored. Modern farming practices subject the moisture-laden corn to dryers using LP (liquid petroleum) gas heaters. Earlier farmers traditionally used ventilated cribs where the corn was naturally dried by the sun's energy.

Select a number of acres (any number will do), and assume that all of this acreage is planted in corn. Postulate an estimated yield for your acreage (bushel/acre). It takes one gallon of LP gas to dry every six bushels of corn.

Corn drying by itself in the field could increase losses as much as 5% of the total crop. Based on the current cost of LP gas (check with your local dealer) and the current market value for a bushel of corn (most daily papers will have a daily quotation for corn), figure out how much the traditional process of natural solar dehydration would cost you. How much would LP gas dehydration cost? Which process would be more expensive to the energy budget?

THE SOLAR HYDRO-STILL

It is possible to extract water from soil and plants by using energy from the sun.

Dig a hole 25 to 30 cm deep and 75 to 85 cm in diameter. Place the removed soil aside. Place a collection container in the center of the deepest part of the bottom of the bowl-shaped hole. Stretch a strong, transparent plastic sheet across the top of the hole and anchor it in several places with small rocks. Place a weighted object, like a rock, in the middle of the plastic sheet directly over the collection container. The weighted object should depress the plastic sheet. This depression should form at least a 34 degree angle from the horizontal plane. Using the previously removed soil, seal the edge of the plastic with soil and small rocks.

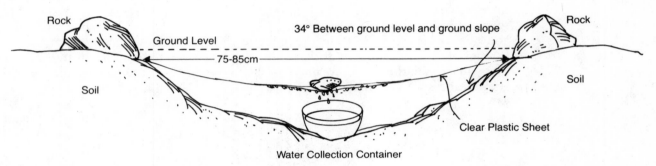

Water Collection Container

After thirty minutes of sunlight, carefully remove the plastic sheet and remove the collection container. Pour the contents into a graduated cylinder and record the amount of water collected in milliliters.

Repeat the process, but this time fill the bottom of the hole with 1 to 3 inches of plant cuttings, cut grass, and so forth. Position the collection container in the center of the hole and cover the hole up with the plastic sheet. After another thirty minutes of direct sunlight, remove the collection container and record the amount of water collected in milliliters.

- Where do you think the water comes from?
- Why does the plastic sheet need to be inclined approximately 34 degrees?
- Explain how the solar hydro-still operates.
- Is the collected water sample hard or soft water?
- How might this device someday save your life?

SOLAR COOKING

Food can be prepared utilizing a solar cooker. Solar cookers are very easy to make. Find an empty cylindrical, oatmeal container and glue the cover back on (use a permanent glue). Locate the center of the top and bottom of the cylinder (remember, they are circles). One half the diameter of the circle is the radius. Make a mark midway on the radius on both circles. This point should be in the same location on each circle. This point will be the focal point of your solar cooker.

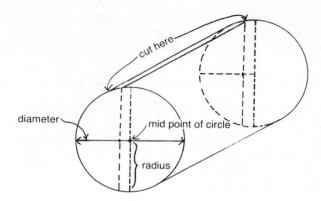

Cut the cylindrical oatmeal container in half, lengthwise. Do not cut the half that you've marked your focal point on. Line the smaller segment of the cylinder with aluminum foil. Glue the foil to the container with rubber cement. Get some bamboo fondue skewers (you can buy them in the appliance department of a local department store or you can use unpainted coathangers, but they get very hot) and pass these through each focal point. Set the cylindrical segment inside another support frame. (This frame can be half of a shoe box.) You need to cut two holes in the sides of the frame for the skewers to fit through. Put your cooker in direct sunlight. The cooker should be adjusted so that you can see the sun's rays cross on the focal points. Put some food on the skewer and start cooking!

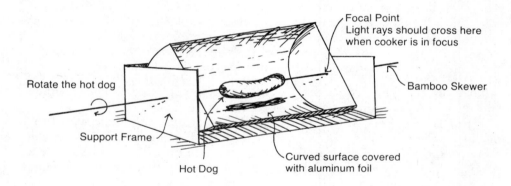

Tomorrow: Which Way?

Our life style was never supposed to end. All the elements of the American way of life were supposed to get bigger, better, go faster, be heavier, longer in length, occur more often, and with greater vitality. We wanted more of everything, all the time. The only outcry that might have been heard was, "What do we do for an encore?" Inexpensive, readily abundant energy was the axle that all this "progress" revolved around.

The energy crisis was inevitable. Our non-renewable resources are finite. They are depletable regardless of the enormity of the original deposits. So now that we know we're running out of our fossil fuels, what do we do?

The energy crisis has raised many questions to which there are no immediate, simple answers. It is difficult to predict what the world will be like after we deal with problems arising from the energy crisis and turn to developing alternative energy sources. The results of positive reactions to the energy crisis could include population stabilization, precise control of our resources for the maximum good of all the people and the minimum impact on the environment, and a highly disciplined society that is educated to achieve pre-set goals.

DID YOU KNOW?

Lumber operations waste from 20 to 60% of the trees they process. The United States has huge forests of fast-growing scrub trees that can be cut into small wood particles. Oil can be derived from these chips. A barrel of oil can be processed from 250 kilograms of wood chips. Other biomass resources, such as municipal sewage and organic waste from household garbage, can supply up to 10% of the present national consumption of energy.

What Can We Do?

Judging from man's creativity in history, we will find suitable, additional energy sources. Our search for them may not follow a timetable that is compatible with our immediate desires, but the end result will be our adjustment to energy sources that do not damage the quality of the environment, and to a life style where energy resources are conserved and distributed for the maximum good of society.

Nature strives continuously for balance. Our past practices of neglecting how we complemented the cycle of life has caused us to wallow in a variety of problems. For example, man's waste problems cannot be solved by dumping waste into the oceans or burying them underground. Nature's waste, by contrast, more appropriately fits into the life cycle. Dead, decaying materials in the forest contribute to enriching the soil and thus contribute to the life cycle. Man, in his constant strife to conquer nature, may learn only his place in nature and how to survive within the well-defined parameters of life. This does not imply a "roll over and play dead" society, but rather an ingenuous society that utilizes new ideas and technologies that are compatible with the environment and the well-being of all people. This will be accompanied by a switch from a highly self-indulgent and materialistic society to a more disciplined, cooperative, functionally oriented society. Society will learn to exist within an energy budget. The throw-away society will disappear. All materials used in the processing of any one object will be reused indefinitely. Junkyards may well be renamed "Resource Rehabilitation Centers." Untapped resources will serve only as reserves from which replacement amounts will be drawn to make up manufacturing production losses and to allow for planned, required increases in productivity.

DID YOU KNOW?

It is predicted that by the year 2000, American farmers might be growing oil as a major cash crop. Oil can be produced during photosynthesis from a poinsettia-like bush, euphorbia lathyris.

The disequilibrium situation brought about by the energy crisis will be brought into balance as alternative energy sources are accepted. This transition will not be made without outcries from opponents, failures in accomplishment, and a continued yearning to return to the "good old days." We believe that the result will be the emergence of a society that is educated, informed, and stronger mentally and physically. Life will go on; perhaps not without discomfort during the adjustment process, but in the long run, we may all benefit more than we realize. A creative society accompanied by leadership that rises to the occasion will enable us to anticipate a desirable future.

Sources for Additional Energy Education Ideas and Activities

The ABC's of Electricity. Channing L. Bete Company, Inc., Greenfield, MA 01301, 1975, 15 pp. $.50

American Petroleum Institute, 1801 K Street, N.W. 1, Washington, D.C. 20006. Free materials

Answers to 40 of the Most Commonly Asked Questions About Electricity. General Electric, Advertising and Promotions Dept., 705 Corporation Park, Scotia, N.Y. 12302, 12 pp. Free materials

Award Winning Energy Education Activities for Elementary and High School Teachers. U.S. Department of Energy, Technical Information Office, P.O. Box 62, Oak Ridge, TN 37830, 38 pp.

The Best Present of All. Oliver A. Houch, reprinted from *Ranger Rick Nature Magazine,* April, 1974. National Wildlife Federation, 1412 Sixteenth Street, N.W., Washington, D.C. 20036, Single copy free, additional copies $.30 each.

Bringing Energy to the People: Washington, D.C. and Ghana (Grades 6—7). U.S. Department of Energy, Technical Information Office, P.O. Box 62, Oak Ridge, TN 37830, 1978, 63 pp. Free materials

Canada, Mexico, and the United States: Energy Mix or Mix-up? Department of Energy, P.O. Box 62, Oak Ridge, TN 37830, 1979. An energy unit for grades 6—9 that discusses the relationships of energy development among the three countries of North America.

A Century of Light. Thomas Alva Edison Foundation, Cambridge Office Plaza, Suite 143, 18280 West Ten Mile Road, Southfield, MI 48075, 1979, $2.00 each plus .50 postage. This is an activity list for Grades K—3 or 4—6 dealing with light.

Coal: Answers to Your Questions. No. 78-33, Edison Electric Institute, Public Information, 90 Park Avenue, N.Y., N.Y. 10016. Free materials

Community Workers and the Energy They Use (Grade 2). U.S. Department of Energy, Technical Information Office, P.O. Box 62, Oak Ridge, TN 37830, 1976, 61 pp. Free materials

The Complete Indiana Energy Saving Home Improvement Guide. Indiana Department of Commerce, Energy Group, Consolidated Building, 115 N. Pennsylvania Street., Indianapolis, IN 46204. Free materials

Computations About the Sources and Conservation of Energy. Governor's Energy Office, State of Florida, 301 Bryant Bldg., Tallahassee, FL 32304.

Consumer Pamphlets on Home Appliances and Energy. Association of Home Appliance Manufacturers, 20 North Wacker Drive, Chicago, IL 60606.

Dialing Down the Energy Crises. Gerald H. Krockover, *Science and Children,* Vol. 13, No. 2, October, 1975, pp. 18—20.

An Educator's Guide to the Three E's: Energy/Ecology/Economics. Consumer Information Services, Sears, Roebuck and Company, D1703, Sears Tower, Chicago, IL 60684, 1978, 19 pp. Free materials

"E" The Magnificent Magician. Department of Energy, Technical Information Office, P.O. Box 62, Oak Ridge, TN 37830, 1979. A kindergarten energy unit that uses a hand puppet named "E" (energy).

The Economy of Energy Conservation in Educational Facilities. Educational Facilities Laboratories, Inc., 850 3rd Avenue, New York, N.Y. 10022, 1978, 81 pp. $4.00

Electricity and Our Energy Future. Edison Electric Institute, Public Information, 90 Park Avenue, New York, N.Y. 10016, 16 pp. Free materials

Electricity, the Individual and the Energy Crisis. #84522, J. C. Penney Company, Inc., Educational Relations Dept., 1301 Avenue of the Americas, New York, N.Y. 10019. $1.25

Encyclopedia of Energy. Daniel N. Kapedes, Editor-in-Chief, McGraw-Hill Book Company, 1221 Avenue of the Americas, New York, N.Y. 10020, 785 pp. $24.50

Energy. National Wildlife Federation, 1412 Sixteenth Street, N.W., Washington, D.C. 20036.

Energy. Scientific American, W. H. Freeman and Company, 660 Market Street, San Francisco, CA 94104, 1979, 221 pp.

Energy Access. Maine Audubon Society, Energy Department, 118 U.S. Route One, Falmouth, ME 04105. $2.00

Energy Activities for Junior High Social Studies. Minnesota Department of Education, 640 Capital Square Bldg., 550 Cedar Street, St. Paul, MN 55101, 1977. Free materials

Energy Activities for the Classroom. Herbert Coon and Michelle Alexander, ERIC/SMEAC Information Reference Center, 1200 Chambers Road, Columbus, OH 43212, 1976, 148 pp. $4.50

Energy and Conservation Education: Activities for the Classroom (Grades 4–6). *Energy and Man's Environment.* QA 224 S.W. Hamilton, Suite 301, Portland, OR 97201, 1977. $25.00

An Energy Curriculum for the Middle Grades. Indiana Department of Public Instruction, Division of Curriculum, Room 229, State House, Indianapolis, IN 46204, 1979, 93 pp.

Energy and Ecology: Are You Involved? Youth Education, Inc., Public Education Association, 20 West 40th Street, New York, N.Y. 10018, 1974, 56 pp.

Energy and Matter. Instructor Curriculum Materials, 757 Third Avenue, New York, N.Y. 10017. Six-poster set, $4.95 pre-paid.

Energy and Conservation of Renewable Resources. Soil Conservation Society of America, 7515 Northeast Ankeny Road, Ankeny, IA 50021. Free materials

Energy and Education Newsletter. National Science Teachers' Association, 1742 Connecticut Avenue, N.W., Washington, D.C. 20009. Free materials

Energy and My Environment, K–6 Teachers Guide. Governor's Energy Office, State of Florida, 301 Bryant Bldg., Tallahassee, FL 32304.

Energy and Our Environment. Corporate Communications Dept., Union Oil Company of California, Box 7600, Los Angeles, CA 90051. Free materials

Energy and Society: Investigations in Decision Making. Biological Sciences Curriculum Study (BSCS), Hubbard Scientific Company, Box 104, Northbrook, IL 60062. Complete program, $85.00

Energy and Transportation (Grade 3). U.S. Department of Energy, Technical Information Office, P.O. Box 62, Oak Ridge, TN 37830, 1967, 61 pp. Free materials

Energy and You. Topeka Outdoor Environmental Education Center, 1601 Van Buren, Topeka, KS 66612. $2.50

The Energy Book. Gerald H. Krockover, in *Instructor,* October, 1978, pp. 57–65. Available from *Instructor,* 757 Third Avenue, New York, N.Y. 10017.

The Energy Challenge (Grades 5–8). U.S. Department of Energy, Technical Information Center, P.O. Box 62, Oak Ridge, TN 37830, 1976. Free materials

Energy Choices For Now: Saving, Using, Renewing. National Education Association, Washington, D.C. 20020. Stock No. 381-12402

Energy Conservation Experiments You Can Do. Thomas Alva Edison Foundation, Cambridge Office Plaza, Suite 143, 8280 West Ten Mile Road, Southfield, MI 48075, 1978, 32 pp. Free materials

Energy Conservation: Understanding and Activities for Young People. Superintendent of Documents, U.S. Government Printing Office, Washington, D.C. 20402. Stock #041-018-00091-7, 1975, 20 pp. $.85

The Energy Crisis. Channing L. Bete Company, Inc., Greenfield, MA 01301, 1974, 15 pp. $1.50

The Energy Crisis in the Public Schools: Alternative Solutions. Ventura County Superintendent of Schools, County Office Building, Ventura, CA 93001, 78 pp. $2.00

The Energy Crisis — What You Can Do About It. Amoco Teaching Aids, P.O. Box 1400 K, Dayton, OH, 45414. $1.00

An Energy Curriculum For the Elementary Grades. Unit I: Energy and You (K–1); Unit II: Energy and Your Community (2–3); Unit III: Energy in Action (4–6). Department of Public Instruction, State of Indiana, Division of Curriculum, Room 229, State House, Indianapolis, IN 46204, 413 pp. Free materials

Energy Education and the School Curriculum. National School Boards Association Research Report, 1055 Thomas Jefferson Street, N.W., Washington, D.C. 20007, 1980, 35 pp. $7.50

Energy Education Assessment Materials. National Assessment of Educational Progress, Suite 1700, 1860 Lincoln Street, Denver, CO 80295.

Energy Education Film Catalog. United States Dept. of Energy, P.O. Box 62, Oak Ridge, TN 37830. Free materials

Energy Education Publications List. U.S. Department of Energy, P.O. Box 62, Oak Ridge, TN 37830. Free materials

Energy/Environment Fact Book. U.S. Environmental Protection Agency, Research and Technical Information Staff, Cincinnati, OH 45268. Free materials

Energy-Environment Mini-Unit Guide. Stephen M. Smith, Editor, National Science Teachers Association, 1742 Connecticut Avenue, N.W., Washington, D.C. 20009, 1975, 220 pp.

Energy Fact Sheets. League of Women Voters Education Fund, 1730 M Street., N.W., Washington, D.C. 20036, 2 pp. each. $.20 per copy, one set of all eleven issues, $1.00

Energy From the Sun. Melvin Berger, Thomas Y. Crowell Company, 10 E. 53rd Street, New York, N.Y. 10022, 1976, 34 pp.

Energy: How and Where Should We Get It? From Coal? From Nuclear Power? How Do They Compare? Vital Issues, Vol. 26, No. 10, Center for Information on America, Washington, CN 06793, 1977, 4 pp. $.45

Energy Kits. Amoco Teaching Aids, P.O. Box 1400 K, Dayton, OH 45414. *The Energy Crisis*, $1.00 and *Living with Energy*, $1.00

Energy: Knowledge and Attitudes. Report No. 08-E-01, National Assessment of Educational Progress, Education Commission of the States, Suite 700, 1860 Lincoln Street, Denver, CO 80295, 46 pp. December, 1978. $3.75

Energy 1978 — Where America Stands on the Energy Crisis. Visual Products Division, 3M, Center, St. Paul, MN 55101, 20 pp. Free materials

Energy Packet. Information Services, Sierra Club, 530 Bush St., San Francisco, CA 94108. Set of 6 articles, $.50

The Energy Problem. SC-4300, World Book-Childcraft International, Inc., Merchandise Mart Plaza, Chicago, IL 60654, 8 pp, 1976. $.20

Energy Puzzles. William J. Crouch, Hayes School Publishing Company, Inc., Wilkinsburg, PA, 1975, 18 pp. $2.50

Energy Resources: How Much Left? Conservation Consultants, 417 Thorn Street, Sewickley, PA 15143, 1979. An energy unit for Grades 4—9. $1.00 plus $1.00 postage

Energy: Selected Resource Materials for Developing Energy Education/Conservation Programs. National Wildlife Federation, 1412-16th Street, N.W., Washington, D.C. 20036, 1978, 36 pp. An excellent source for additional energy education materials. Free materials

Energy: The Thread of Life. Department of Energy, P.O. Box 62, Oak Ridge, TN 37830, 1979. An energy unit for grades 3—5 that examines the dependence of ecological systems on energy.

Energy in Transition: 1985—2010. National Academy of Sciences, W. H. Freeman Co., 660 Market St., San Francisco, CA 94104, 1980, 675 pp. $10.95

The Energy We Use (Grade 1). U.S. Department of Energy, Technical Information Center, P.O. Box 62, Oak Ridge, TN 37830, 1977, 42 pp. Free materials

Energy: Why We Must Conserve It Now. Edison Electric Institute, Public Information, 90 Park Avenue, New York, N.Y. 10016, 16 pp. 1977. Free materials

Environmental Education, Energy-Society (Grades 4—12). Bureau of Secondary and Elementary Education, DHEW/OE, Washington, D.C. 20009, $3.50

Environmental Education, Energy-Transportation (Grades K—8). The New Jersey State Council for Environmental Education, Montclair State College, Montclair, NJ 07042.

Everyday Conservation: Energy and Resources. Instructor manual available from Social Studies School Services, 10000 Culver Blvd. P.O. Box 802, Dept. E., Culver City, CA 90230. Order #INS762, 15 color posters plus 16 task cards. $5.95

Factsheet Series. National Science Teachers Association, 1742 Connecticut Avenue, N.W. Washington, D.C. 20009, 4 pp. each. Free materials

Facts About Oil. American Petroleum Institute, Publications and Distribution Section, 2101 L Street, N.W., Washington, D.C. 20037. $.35 each

Geothermal Energy. Corporate Communications Dept. Union Oil Company of California, Box 7600, Los Angeles, CA 90051. Free materials

Home Sweet Earth. Marie Meaney, Highline Public Schools, Seattle, Washington. Available through the Educational Resource Information Center (ERIC), 1200 Chambers Road, Columbus, OH 43212. For first grade, ED 132 009.

Household Energy. Individualized Science Instructional System, Ginn and Company, Lexington, MA 02173, 1976, 76 pp.

The Household Energy Game. Communications Office, University of Wisconsin, Sea Grant College Program, 1800 University Avenue, Madison, WI 53906, 20 pp., 1974. Free materials

Idaho Energy Education Materials. Idaho Office of Energy, Statehouse, Boise, ID 83720. Free materials

Industrial Energy Conservation: 101 Ideas At Work. Energy Management Section, General Motors Corp., 3044 W. Grand Blvd., Detroit, MI 48202. Free materials

Iowa Energy Conservation Packet (one each for Grades K—6). Iowa Energy Policy Council, 215 East Seventh Street, Des Moines, IA 50319, 1977. $2.00 each

It's Your Environment. Sherry Koehler, Ed., Charles Scribner and Son, 597 Fifth Avenue, New York, N.Y. 10017, 1976, 218 pp. $1.95

Learning About Energy (Grades K–3). David C. Cook Publishing Company, School Products Division, Elgin, IL 60120, 1978. Ten full color photographs plus 24 pages teacher's manual. $4.50

Learning Activity Packets. Energy Management Center, P.O. Box 190, Port Richey, FL 33568. Three booklets, 55 pp. each. $1.75 for three. Teacher's guide, $8.25

Learning Activity Posters Energy For Today and Tomorrow. January 1978, Starting Points, Dept. 3705, Box 818, Maple Plain, MN 55348, $2.00

Light and Heat Energy. Conservation Consultant, 417 Thorn Street, Sewickley, PA 15143, 1979. An energy unit for grades K–3. $1.00 plus $1.00 postage

Living With Energy. Youth and Educational Activities, Standard Oil Company (Indiana), Mail Code 3705, 200 E. Randolph Drive, Chicago, IL 60601. Free materials

Looking for Energy? A Guide to Information Resources. American Petroleum Institute, 2101 L Street, N.W., Washington, D.C. 20037, 17 pp. Free materials

The Maine Teacher's Energy Primer (Grades 5–7). Maine Audubon Society, Energy Dept., 118 U.S. Route One, Falmouth, ME 04105. $5.00

Map of Coal Areas in the United States. National Coal Association, Coal Building, 1130 17th St., N.W., Washington, D.C. 20036, 1977. Free materials

Mickey Mouse and Goofy Explore Energy and *Mickey Mouse and Goofy Explore Energy Conservation.* Public Affairs Dept., Exxon, U.S.A., P.O. Box 2180, Houston, TX 77001. Free materials

Mineral Resources. Bureau of Mines, Publications Branch, 4800 Forbes Avenue, Pittsburgh, PA 15213. Coal Products Tree and The Petroleum Tree. Free materials

The Natural Laws of Energy. Conservation Consultants, 417 Thorn Street, Sewickley, PA 15143, 1979. An energy unit for grades K–3. $1.00 plus $1.00 postage

Networks: How Energy Links People, Goods, and Services (Grades 4–5). U.S. Department of Energy, Technical Information Office, P.O. Box 62, Oak Ridge, TN 37830, 1978. 102 pp. Free materials

NRG. Public Relations Dept., American Petroleum Institute, 2101 L Street, N.W., Washington, D.C. 20037, 16 pp. Illustrated storybook about energy.

Nuclear Power. Atomic Industrial Forum, Inc., Public Affairs and Information Program, 7101 Wisconsin Avenue, Washington, D.C. 20014, 9 booklets. $.08 each

Oil. Amoco Teaching Aids, Mail Code 3705, P.O. Box 5910A, Chicago, IL 60680. "Catalysts and Crude," 22 pp., "Oil Depth," 30 pp., "Our Energy Future," 5 pp. Free materials

Oil. Phillips Petroleum Co., Corporate Writing - 4D4PB, Bartlesville, OK 74004. "The ABC's of Oil and the Petroleum Industry," 16 pp., "The Magic World of Petrochemicals," 33 pp., "The Story of Oil and Gas." Free materials

100 Ways to Save Energy and Money in the House. Office of Energy Conservation, Dept. of Energy, Mines, and Resources, 580 Booth St., Ottawa, CAN K1A OE4, 160 pp.

Only Silly People Waste (Primary Grades). Norah Smaridge, Abingdon Press, Nashville, TN, 1976. $5.50

Perspectives on Energy: Issues, Ideas and Environmental Dilemmas. Lon C. Ruedisili and Morris W. Firebaugh, Eds. Oxford University Press, New York, N.Y. 10001, 1978, 591 pp.

Policy for Energy Education, Grades Kindergarten through 12. Education Commission of the States, Suite 300, 1860 Lincoln St., Denver, CO 80295, 1979. Free materials

Publications List. National Solar Heating and Cooling Information Center, P.O. Box 1607, Rockville, MD 20850. Free materials

The Real Cost of a Refrigerator. Conservation Consultants, 417 Thorn St., Sewickley, PA 15143, 1979. Activity unit for grades 4–9. $1.00 plus $1.00 postage

Recycle? In Search of New Policies for Resource Recovery. No. 132, League of Women Voters, Education Fund, 1730 M Street, N.W., Washington, D.C. 20036, 39 pp. $.75

Recycling Scrap Metal. Institute of Scrap and Iron and Steel, Inc., 1627 K Street., N.W., Washington, D.C. 20006. Background kit for educators. Free materials

Save Energy and Have Fun with Textiles. American Textile Manufacturers Institute, 1101 Connecticut Ave., N.W., Washington, D.C. 20036, 36 pp. Free materials

Save It! Keep It! Use It Again! R.J. Lefkowitz, Parent's Magazine Press, New York, N.Y., 1977, 64 pp.

Science Activities In Energy (Grades 4–6). U.S. Department of Energy, Technical Information Office, P.O. Box 62, Oak Ridge, TN 37830. Free materials

Selected Department of Energy Publications List. U.S. Department of Energy, Technical Information Office, P.O. Box 62, Oak Ridge, TN 37830. Free materials

Selected U.S. Government Publications. S.L. Mail List, Washington, D.C. 20402. Ask to be placed on the mailing list for the free bi-weekly list of selected U.S. Government Publications.

Seven Edison Experiment Booklets. David Schantz, Curator, Charles Edison Fund, 101 Harrison St., East Orange, NJ 07101. $.50

Shell Answer Books. P.O. Box 61609, Houston TX 77208.

Solar Energy. William W. Eaton. U.S. Department of Energy, Technical Information Office, P.O. Box 62, Oak Ridge, TN 37830, No.-74-600179, 49 pp.

Solar Energy Classroom Reality. Thomas Geier and James R. Wailes. *Science and Children,* Vol. 17, No. 1, September, 1979, pp. 19–20.

Solar Energy Education Packet for Elementary and Secondary Students. Center for Renewable Resources, 1028 Connecticut Ave., N.W., Washington, D.C. 20036.

Solar Energy Project Materials. Public Documents Distribution Center, Dept. 17, Pueblo, CO 81009, 1979. Solar Energy Project Teacher's Guide (S/N 061-000-00234-1) 84W9. $2.20; General Solar Topics 95W9, $2.50; Junior High Science 86W9 (S/N 061-000-00228-6) $2.75; Earth Science 87W9 (S/N 061-000-00232-4) $2.75; Chemistry and Physics 88W9 (S/N 061-000-00229-4) $2.20; Biology 89W9 (S/N 061-000-00230) $1.70; Text 90W9 (S/N 061-000-00232-2) $2.75; Reader 91W0 (S/N 061-000-00235-9) $2.75

Solar Hot Water and Your Home. National Solar Heating and Cooling Information Center, P.O. Box 1607, Rockville, MD 20850, 20 pp.

Special Report on Energy, National Geographic Society, Department 5000, Washington, D.C., 20036, February, 1981, 115 pp. $1.45

The Status of Energy Education Policy. Education Commission of the States, Suite 300, 1860 Lincoln St., Denver, CO 80295, 1979. $4.10

Tapping Earth's Heat. Patricia Lauber, Garrard Publishing Co., Champaign, IL, 1978, 64 pp.

Teaching About Energy Awareness: 33 Activities. Materials Distribution Center for Teaching International Relations, University of Denver, CO 80208, 1978, 1979 pp. $7.95

Tilly's Catch—A Sunbeam Coloring Book. Solar Service Corporation, 306 Cranford Road, Cherry Hill, NJ 08003, 32 pp. $1.50

Tips for Energy Savers. Consumer Information Center, Pueblo, CO 81009, 46 pp. Free materials

The Trans-Alaska Pipeline. Alyeska Pipeline Service Co., 1835 South Bragaw St., Anchorage, AL 99504, 15 pp. Free materials

Trees and Forest Management. 5B-086, U.S. Government Printing Office, Washington, D.C. 26402, 12 pp. Free materials

Two Energy Gulfs. Department of Energy, Technical Information Center, P.O. Box 62, Oak Ridge, TN 37830. An energy unit for grades 6–7. Free materials

200 Ways to Save on Energy in the Home. Acropolis Books, Colortone Bldg., 2400 17th St., N.W., Washington, D.C. 20009, 105 pp., 1978. $4.95

Underground Furnaces: The Story of Geothermal Energy. Irene Kiefer, William Morrow & Co., 105 Madison Ave., New York, N.Y. 10016, 1976, 63 pp.

The United States and World Energy. No. 8904, Bureau of Public Affairs, Office of Media Services, U.S. Dept. of State, Washington, D.C., 1977, 39 pp. Free materials

Visual Masters in Energy Resources: Past, Present, and Future. J. Weston Walch, P.O. Box 658, Portland, MA 04104. Order #L6409R-4. $9.00/set

What Is a Gas? Educational Services, American Gas Association, 1515 Wilson Blvd., Arlington, VA 22209. Free materials

Wind Is to Feel. Shirley Cook Hatch. Coward, McCann & Geoghegan, Inc., 200 Madison Ave., New York, N.Y. 10016, 1973, 30 pp.

Wind Machines. 33X9 (S/N 038-000-00272-4) Public Documents Distribution Center, Department 18, Pueblo, CO 81009, 1976, 77 pp. $2.25

Your Energy World (Grades K–6). U.S. Department of Energy, Technical Information Center, P.O. Box 62, Oak Ridge, TN 37830. 8 units plus poster. Free materials

Additional Information Concerning Electric Vehicles

B & Z Electric Car, 3346 Olive Ave., Signal Hill, Calif. 90806

Battery Power Unit Corp., Rt. 3, Box 700, Golden, Colo. 80401

Die Mesh Corp., 629 Fifth Ave., Pelham, N.Y. 10803

Electric Fuel Propulsion Corp., Robbins Executive Park East, 2191 Elliott Ave., Troy, Mich. 48084

Electric Auto Corp., 2237 Elliot Ave., Troy, Mich. 48084

Electric Passenger Cars Inc., 5127 Galt Way, San Diego, Calif. 92117

Electric Vehicle Associates Inc., 9100 Bank St., Valley View, Ohio 44125

General Engines Co. Inc., 591 Mantua Blvd., Sewell, NJ 08080

Globe-Union Inc., 5757 North Greenbay Ave., Milwaukee, Wis. 53201

H-M Vehicles Inc., 6276 Greenleaf Terrace, Apple Valley, Minn. 55124
Hummer Inc., Box 2099, Wichita, Kans. 67201
Huber Engineering Co., Box 17, Galva, Ill. 61434
Hybricon Inc., 11489 Chandler Blvd., North Hollywood, Calif. 91601
Jet Industries Inc., 4201 South Congress, Austin, Tex. 78745
JMJ Electronics Corp., 4415 Highline Blvd., Oklahoma City, Okla. 73125
Kaylor Energy Products, 1918 Menalto Ave., Menlo Park, Calif. 94025
Lyman Metal Products, 15 Meadow, South Norwalk, Conn. 06856
Marathon Electric Vehicles Ltd., 8305 Le Creusot St., Montreal, H1P 2A2, Quebec, Canada
Palmer Industries, Box 707, Union Station, Endicott, N.Y. 13760
Palmer Sales & Service Inc., 3042 West Colter St., Phoenix, Ariz. 85017
Quincy-Lynn Enterprises Inc., Box 26081, Phoenix, Ariz. 85020
Real Electric Vehicles, 727 N.E. Fifth Ave., Gainesville, Fla. 32601
South Coast Technology, Inc., 5553 Hollister Ave., Goleta, Calif. 93017
3-E Vehicles, Box 19409, San Diego, Calif. 92118
U.S. Electricar Corp., White Pond Rd., Athol, Mass. 01331

Index

W

*2284-5
1982
5-01
C